Climate Change and Adaptive Innovation

The world is witnessing climate change. As responsible citizens of planet earth, we can actively participate in the co-creation of actionable knowledge and solutions. There may not be a single and linear pathway to adaptation anymore. This book explores multiple and iterative pathways of adapting to climate change and its impacts.

Climate Change and Adaptive Innovation introduces an adaptive innovation model that has its premise on core values of justice, care and solidarity. Navigating collectively through shared conversations and dialogic processes, this model showcases how we could embark on an enduring journey where diverse actors could collaboratively make informed choices and take necessary actions to enhance the safety and security of their lived environment. Rooted in action research, it is envisaged that this model could enable us to facilitate the designing and implementation of people-centred ethical adaptation projects.

This book will be of interest to social workers, social scientists and development practitioners who are engaged in the field of climate justice, adaptation, social innovation and sustainable livelihoods. Social work educators and students will certainly draw inspiration from the stories that are shared in this book. It will further motivate many transdisciplinary professionals to engage with action research as a method of innovation, reflection and practice.

Sunil D. Santha is presently working at the Centre for Livelihoods and Social Innovation, School of Social Work, TISS, India. He has associated with natural resource dependent communities for several years. As an expert in Environmental Social Work, supplemented by his interests in Sociology of Knowledge and Systems Thinking, he has keen interests in the field of environmental risks, climate justice and sustainable development. He believes in action research and reflective practice as creative pathways to design participatory models of social innovation and emergent livelihoods.

Indigenous and Environmental Social Work Series

Series Editor: Hilary Weaver, University as Buffalo, USA

Sustainability is the social justice issue of the century. This series adopts a global and interdisciplinary approach to explore the impact of the harmful relationship between humans and the environment in relation to social work practice and theory. It offers cutting-edge analysis, pioneering case studies and current theoretical perspectives concerning the examination and treatment of social justice issues created by a disregard for non-Western cultures and environmental detachment. These books will be invaluable to students, researchers and practitioners in a world where environmental exploitation and an ignorance of indigenous peoples are violating the principles of social justice.

Titles

Decolonised and Developmental Social Work
A Model from Nepal
Raj Kumar Yadav

Eco-activism and Social Work
New Directions in Leadership and Group Work
Edited by Dyann Ross, Martin Brueckner, Marilyn Palmer and Wallea Eaglehawk

Climate Change and Adaptive Innovation
A Model for Social Work Practice
Sunil D. Santha

For a full list of titles in this series, please visit: www.routledge.com/Indigenous-and-Environmental-Social-Work/book-series/IESW

Climate Change and Adaptive Innovation

A Model for Social Work Practice

Sunil D. Santha

Routledge
Taylor & Francis Group

LONDON AND NEW YORK

First published 2020 by Routledge

2 Park Square, Milton Park, Abingdon, Oxon OX14 4RN
605 Third Avenue, New York, NY 10017

Routledge is an imprint of the Taylor & Francis Group, an informa business

First issued in paperback 2022

Publisher's Note

The publisher has gone to great lengths to ensure the quality of this reprint but points out that some imperfections in the original copies may be apparent.

British Library Cataloguing-in-Publication Data
A catalogue record for this book is available from the British Library

Library of Congress Cataloging-in-Publication Data
Names: Santha, Sunil D., author.
Title: Climate change and adaptive innovation : a model for social work practice / Sunil D. Santha.
Description: 1 Edition. | New York : Routledge, 2020. |
Series: Indigenous and environmental social work | Includes bibliographical references and index.
Identifiers: LCCN 2019055709 (print) | LCCN 2019055710 (ebook) |
ISBN 9780367195557 (hardback) | ISBN 9780429203138 (ebook)
Subjects: LCSH: Climatic changes–Social aspects. | Climatic changes–Economic aspects.
Classification: LCC QC903 .S26 2020 (print) | LCC QC903 (ebook) |
DDC 363.738/74532–dc23
LC record available at https://lccn.loc.gov/2019055709
LC ebook record available at https://lccn.loc.gov/2019055710

ISBN: 978-0-367-19555-7 (hbk)
ISBN: 978-1-03-233662-6 (pbk)
DOI: 10.4324/9780429203138

Typeset in Times New Roman
by Integra Software Services Pvt. Ltd.

A tribute to
Mahatma Gandhi
on his 150th Birth Anniversary

Contents

Figures

Stories from the field

Acknowledgements

Oh, Mother! I bow to Thee again and again,
The embodiment of bliss; and the slayer of worldly sufferings;
The source of wisdom, care and devotion.
My parents and teachers – You helped me to explore this world,
With care and responsibility.
Bhuvan, Sree and *Swami,*
Each page in this book has an inspiration from you,
And a story to remember.
My *friends, colleagues* and *students,*
Thanks for being there for me! Your presence mattered the most.
The Tata Institute of Social Sciences,
For inculcating the spirit of social innovation and transformative change.
The true *innovators at the grassroots,*
I came as an outsider, you took me in as one of yours,
Your shared conversations made this pathway possible.

***

Abbreviations

ACF	Action Contre la Faim (Action Against Hunger)
BUIOH	Baylor University Institute for Oral History
COMEST	World Commission on the Ethics of Scientific Knowledge and Technology
CSR	Corporate Social Responsibility
DFID	Department for International Development
FAO	Food and Agriculture Organisation
GIS	Geographic Information Systems
ICIPE	International Centre of Insect Physiology and Ecology
ICRISAT	International Crops Research Institute for the Semi-Arid Tropics
IFAD	International Fund for Agricultural Development
IIED	International Institute of Environment and Development
ILO	International Labour Organisation
IPCC	Intergovernmental Panel on Climate Change
IRHDP	Integrated Rural Health and Development Project
MSSRF	M.S. Swaminathan Research Foundation
NGO	Non-Governmental Organisation
OECD	Organisation for Economic Cooperation and Development
PAR	Pressure and Release Model
PCVA	Participatory Risk, Capacity and Vulnerability Analysis
PLA	Participatory Learning and Action
PRA	Participatory Rural Appraisal
SFF	Stories from the Field
TGCS	Tree Growers' Cooperative Society
TGMACS	Tree Growers' Mutually Aided Cooperative Society
TISS	Tata Institute of Social Sciences
UN/ISDR	United Nations/International Strategy for Disaster Reduction
UNDP	United Nations Development Programme
VCA	Vulnerability and Capacity Analysis
VDC	Village Development Committee
WHO	World Health Organisation

1 Introduction

As I am writing the prelude for this book, more than 11,000 scientists have co-signed a letter in a reputed journal demanding urgent and necessary action on climate change. I believe that this may be one of the largest gathering of scientists who are explicitly demanding for immediate and proactive climate action. So far, nation states have failed to reduce global carbon emissions to the desired levels and much time has already been wasted in denying climate change. Nevertheless, global surface temperatures are rising, and the increase seems to be at a faster rate than what our scientists had predicted two decades ago. We are witnessing disruptions in weather patterns accompanied by other daunting consequences impacting the balance of diverse ecosystems. The state of oceans, forests, biodiversity, food, water and livelihood security and ultimately our ability to sustain a safe and secure life are in peril.

The media has been reporting extreme environmental events across different parts of the world. Climate models have proved that rising global temperatures and heat waves have accelerated the melting of glaciers at a rate that was not anticipated earlier. In the last few years, many cities across the world have been witnessing higher surface temperatures. Water scarcity in several dryland regions of the world is becoming a new normal. Many countries in Asia and Africa are considerably affected by drought, where farmers are being forced to abandon dying crops; herders are selling their livestock in distress; and women and children are struggling to find enough water to survive. There has been an increase in the loss of human and non-human lives due to extreme weather events. The severe drought in Zimbabwe has been killing elephants and other wildlife.

With rising temperatures, there has also been an increase in the frequency and intensity of cyclonic storms, sea surges and hurricanes. Recently, Mozambique and the Bahamas were devastated by severe cyclonic storms. While I am penning down these opening paragraphs, cyclone Bulbul is fast brewing in the Bay of Bengal. This will be the eighth cyclone that India will be witnessing this year and it will bypass a previous three-decade record of having the maximum number of cyclones in a year. The year 2018 also witnessed seven major cyclones. Further, the cyclone Fani that originated in the Bay of Bengal was one of the strongest storms that the country witnessed in the past 20 years. Fragile ecosystems such as the Chilika Lake and the livelihoods of many poor people dependent on the lake were severely destroyed by the cyclone. The life and livelihood security of several small island communities such as people in the Andaman and Nicobar Islands, the Sundarbans and Lakshadweep region were affected by a series of extreme weather events.

Environmental and humanitarian activists argue that those who contributed least to the wicked problem of climate change are suffering the most. According to them, the world will be staring at a crisis of addressing the basic needs of millions of climate refugees or climate migrants in the coming years. By the year 2050, countries of Sub-Saharan Africa, South Asia and Latin America alone could witness the number of climate migrants to rise up to 143 million (Rigaud et al., 2018). Developed nations are also not free from such challenges. Media reports show that certain coastal towns in United States and Britain are already struggling to relocate communities affected by rising water levels, flooding and submergence of large stretches of land (Davenport and Robertson, 2016; Wall, 2019). And these reports caution us that this is just the beginning. There are other exigencies too. The International Labour Organisation (ILO), for instance, has warned that one of the significant impacts of climate change is going to be unemployment (ILO, 2019). And most of the affected people are going to be the poor and marginalised segments of the population. I commenced the writing of this book with the realisation that we, as social workers and development practitioners could play an important role in enabling vulnerable populations to effectually respond to climate change and co-create opportunities for sustainable development.

The aim of this book is to introduce and elaborate a community-based innovation model for climate change adaptation. I refer to the model as 'adaptive innovation', which has its premise on the core values of justice, care and solidarity. It aims at vulnerability reduction and strengthening the adaptive capacities of community actors in complex social–ecological systems. This model emphasises on people-centred processes by which local community actors collectively can analyse their own situations in the context of social and ecological transitions; forge constructive partnership with other relevant actors to dialogue, ideate and develop working models; and implement and critically observe, reflect and validate their adaptive strategies to the emergent contexts. Navigating collectively through shared conversations and dialogic processes, we could embark on an enduring journey where diverse actors would mutually learn, innovate and make informed choices to enhance the safety and security of their lived environment. Rooted in action research, it is envisaged that the adaptive innovation model could enable us practitioners to facilitate the designing and implementation of people-centred ethical adaptation projects.

The idea of writing such a book had its genesis from my experiences of working with marginalised rural communities in drought affected regions of Andhra Pradesh in India. Later as I travelled across the country, I realised that the vulnerabilities of local communities to climate change and extreme natural hazard events are poignant and they all share similar narratives of struggle and hope. Interactions with practitioners in the field, social work educators and students also showcased a need for elaborating a people-centred practice-based model to stimulate social innovations in climate change adaptation. This book has also drawn substantial inspirations from Gandhian, Feminist and Systems perspectives; and lays its premise on an ethical standpoint of adaptation that is embedded with values of justice, care and solidarity.

Mahatma Gandhi believed in a truthful and non-violent way of life. He believed that human beings should not cause violence to other sentient beings in this earth. I once happened to read about an event in the life of Gandhiji that very much shaped my thinking on sustainability as well. One night, a person in Gandhiji's *ashram* had plucked large number of leaves beyond the quantity that was required. Such a careless approach of dealing with nature pained Gandhiji. According to him, trees were also living beings like humans. And he felt very bad that the leaves were plucked during the resting time of plants and that too beyond what was needed. Gandhiji's perspective of care always insisted that humans nurture a 'living bond' with rest of the non-human world (Gandhi, 1961, p. 303). Gandhiji's celebrated and often-quoted statement that the earth has enough to satisfy the need of all the people, but not for satisfying the greed of some has become very relevant in the context of global warming and climate change.

Gandhiji's philosophy of *Sarvodaya* is based on the principle of well-being of all human as well as other sentient beings. My initial efforts to understand social–ecological systems drew inspiration from similar Gandhian worldviews. He believed in the unity and oneness of all life and its interconnectedness. Many spiritually oriented cultures across the world have alerted us to the significance of this systemic interconnectedness. In his persistent endeavour to hold on to truth (*Satyagraha*), Gandhiji continuously strived to apply his principles in practice, reflect and evolve from these experiences. This approach was his pathway to self-realisation. He strongly believed that non-violence in its true sense has the ultimate solution to any problems affecting this world. Gandhiji envisaged a non-violent economic order based on equality and justice. His notion of solidarity was guided by selfless-service, cooperation and trusteeship. And he believed in dialogic practice that is embedded in learning-by-doing, truthfulness, mutual respect and love. Such a way of life requires mindful engagement, care and compassion with the environment, economy and the spiritual self (Kumarappa, 1957).

My fieldwork experiences and interactions with diverse rural communities has revealed to me the immense significance of local knowledge systems. Community actors are custodians of certain knowledge forms, which enable them to live meaningfully in their local environment, sustain their culture and practice their livelihoods. I also observed that local community actors in the villages and smaller towns often uphold a larger, interconnected systems view of life. They also have an excellent skill in articulating and linking the everyday moments in their life to the larger cosmos. We could see that their random expressions will cover different disciplinary but surprisingly unified or interlinked biological, ecological, social, economic, political, cognitive, philosophical, mythological and spiritual worldviews. Their conversations often do reflect the dynamic and non-linear nature of their relationship with humans and non-humans, their obligations to their ancestors and future generations and to the complex web of life in which all of us belong to. To some extent, they are aware of the larger system of life in which they and us are part, and at the same time they seem to be embedded in their everyday struggles, relationships and responsibilities.

I have observed that for many vulnerable groups, local knowledge is their primary resource to forecast and cope with a natural hazard event. However, in practice there are multiple dominating knowledge systems, and the 'local' is often represented as 'tacit, unscientific and hidden' in the mainstream notions of adaptation. Moreover, the romanticised and functionalist views of local knowledge have its own limitations to unlock the lived realities and aspirations of marginalised and subjugated actors. Further, the complexities involved in addressing a wicked problem like climate change forced me to explore other alternatives of recognising and representing the knowledge of subjugated actors. I discovered that the feminist perspectives of situated knowledge were extremely helpful in designing these alternative pathways. The feminist perspectives that recognise the intersectional, plural and dynamic nature of knowledge systems demands us to be attentive to the contexts of our practice. At a far deeper level, the feminist awareness is based on each actor's experiential knowledge that all life is connected, and both individuals and societies are located within the cyclical processes of nature (Spretnak and Capra, 1984). Such an awareness also finds meaning in nurturing relationships of care embedded in trust and solidarity. Practice based on this awareness would entail us to have shared conversations and dialogic interactions as potential adaptation pathways.

Climate change is a systemic problem and therefore to adapt, we need to look out for systemic solutions. It requires a systemic understanding of the social–ecological context, its dynamics and capabilities to act up on this knowledge. There may not be a single and linear pathway to adaptation anymore. We need to explore multiple and plural pathways. We need to recognise the value of local knowledge, the subjective experiences and situated practices of community actors in collaboratively creating a meaningful synergy out of interconnected, dynamic and complex social–ecological systems. Such a frame of practice requires forging partnerships that recognise and value local needs, aspirations and knowledge of vulnerable groups as the primary base for designing and implementing adaptation projects. My experiences show that action research has tremendous potential to address complex problems at the grassroots level. As committed social workers and as responsible citizens of planet earth, we can actively participate in the co-creation of actionable knowledge and solutions. Such an approach also provides us opportunities to become self-aware and innovatively engage with complex problems of the world.

Enabling vulnerable groups to adapt effectively to variations and uncertainties in the natural environment gains prime importance in our everyday pursuit of social justice, human well-being and empowerment. There is need for us social workers to be critically alert and responsive to the major drifts in climate change adaptation and emergent livelihoods. At the same time, we have to equip ourselves with innovative methodologies in facilitating people-centred adaptation projects. Though the literature has widely discussed the roles that social workers could play as transactional agents in strengthening the adaptive capacities of vulnerable groups, the practice dimensions of these roles are sparsely dealt with. In this book, I attempt to showcase how action research as an approach can be used to bring about transformative change as well as offer immense possibilities for theorisation, co-creation and collaborative learning. The planning and practice of adaptation requires constant attention and reflection on what is being done, what is emerging and what iterations have to be done to move ahead with our actions. The adaptive innovation model that is presented in this book is a step towards this direction. Each phase of innovation can be seen as a process of discovering new ways of participation, co-creation and embedded learning. I am optimistic that the application of this model in specific contexts could demonstrate how diverse adaptation strategies can be iteratively designed taking into account the localised practices, policy frameworks and the relations of power within which they are structured.

The adaptive innovation model is based on certain conceptual foundations. The next chapter of this book introduces some of these key concepts and their interrelationships. It describes the nature of social–ecological systems, their complexities and dynamics in the context of climate change. It also elaborates on the need to define and locate the community as a critical participant actor in climate change adaptation. A larger premise that this book employs towards developing the adaptive innovation model is situated knowledge. Chapter 2 also elucidates the characteristics of situated knowledge in the everyday social encounters of community actors and its relevance in unearthing differentiated strategies of practice. Nevertheless, these practices are also linked to the nature of ideas, imagination, knowledge and institutions that co-exist and emerge in the adaptation landscape. Further, this chapter briefly explains the suitability of action research in developing people-centred climate change adaptation strategies. It also discusses how innovation platforms can be developed as an appropriate institutional medium for facilitating collective action.

Chapter 3 provides an overview of the whole adaptive innovation model. The landscape of adaptive innovation is elucidated by describing the values guiding ethical adaptation, its six practice phases, the embedded cycle of reflective practice and the significance of analysing actor interfaces in adaptation. To begin with, Chapter 3 briefly narrates the values that could guide ethical adaptation, namely climate justice, ethics of care and solidarity. It further presents the six phases of adaptive innovation. This chapter also explains why these phases have to be embedded within the iterative cycles of reflection-for-action, reflection-in-action and reflection-on-action. It also highlights why it is important to analyse the actor interfaces in adaptation to critically engage with the adaptation process and at the same time address the needs of the most vulnerable and subjugated actors.

Chapter 4 elaborates on the important values guiding ethical adaptation to climate change. All adaptation decisions and actions taken could have implications in terms of intersectional, intergenerational and interspecies dimensions. It is in these contexts the adaptive innovation model envisages the presence of three critical elements, namely climate justice, ethics of care and solidarity. Ensuring fairness in response to climate change involves both procedural and distributive justice, where representation of vulnerable groups and recognition of their needs and voices becomes crucial. However, the notion of justice alone would not help in recognising the grounded experiences of marginalised actors to locate and reduce vulnerabilities. This chapter therefore deliberates on the need for an ethics of care and solidarity perspectives as well. Adaptive innovation thus could be understood as a process of transformative change that interweaves diverse pathways of caring solidarity along with the pursuit of justice.

Adaptive innovation is an iterative and reflective process involving diverse actors who are engaged in an interactive frame of analysis, ideation and practice. Chapter 5 of this book explains in detail the six phases of the adaptive innovation model. The six phases are Situational Analysis, Micro-mobilisation, Dialogic Ideation, Action Framing, Piloting and Emergence. The importance of scoping reviews, participatory mapping and analysis of drivers and barriers in situational analysis is explained. The process of micro-mobilisation is narrated with emphasis on shared visioning, participatory risk and vulnerability analysis and formation of innovation platforms. It is envisaged that the members of the innovation platform would engage in dialogic ideation and action framing to imagine diverse pathways of change and build appropriate working models respectively. These models are further piloted for feasibility and scaling up. However, the final phase of emergence is very crucial in adaptive innovation. It highlights those novel and creative phenomena that could arise from our adaptation efforts or else it could represent patterns underlying the anticipated and unintended consequences of our action.

Chapter 6 deliberates upon the various methods of inquiry and practice that could be followed in each of the adaptive innovation phases. Rooted in action research, it illustrates the diverse methods of interaction and data collection techniques that could facilitate ideation, decision-making and collective action in the adaptive innovation processes. Some of these methods that are discussed in this chapter include face-to-face interfaces, historical, visual and participatory narratives and other reflective data sources. Face-to-face interfaces could be different types of interviews, while oral histories could be an enriching source for historical insights and local knowledge. This chapter also briefs on some of the relevant participatory learning and action techniques. The use of photovoice, storytelling, brainstorming workshops, focus groups, social simulations, design charrettes and participatory modelling are also discussed.

Chapter 7 explains the context, method and practice of analysing actor interfaces in adaptive innovation and its embeddedness with reflective practice. It illustrates how the actor interface analysis can be looked up on as a suitable analytical frame to reflect upon crucial social encounters, and on our own understanding and practice. Each social encounter that shapes adaptive innovation could involve diverse social actors with differing interests, knowledge, resources and power. It also corresponds to the differentiated meanings and strategies that diverse actors evolve during the process. It is equally important to reflect upon insider–outsider relationships, our own positionality, perceptions, resources and investments that shape adaptive innovation. Insights from this chapter could also contribute to theory building in the context of climate change adaptation and reflective social work practice.

Chapter 8 presents some of my concluding reflections. Adaptive innovation is an enduring journey where diverse actors navigate collectively to make their lived environment safe and secure, mutually learn and make choices and create meaning out of these experiences. Our methods of inquiry and practice will require continuous reflection, innovation and improvisation that has its primary premise on people-centred knowledge systems and decision-making capabilities. Adaptation decisions have to take into account the social positions of vulnerable and marginalised groups and their situated knowledge-cum-practices. The ultimate outcome of adaptive innovation has to be that all involved actors are able to develop their individual and collective capacities to solve complex adaptation challenges. Spaces that nurture shared conversations and solidarity networks therefore have a crucial role in climate change adaptation.

I have tried to supplement the key discussion of each chapter with stories from the field. I believe as humans, our ability to create meanings from stories will convey the message more effectively and with empathy. And most importantly to me, stories inspire action (Chakraborty, 2018). Most of the stories that I have shared in these chapters are from my own situations of practice. In certain contexts, I have supplemented these anecdotes with stories and narratives shared by journalists and development practitioners. I am hopeful that this book will make meaning to social workers, social scientists and development practitioners who are engaged in the field of climate justice, adaptation and sustainable livelihoods. I also believe that social work educators and students will draw inspirations from the stories that I have shared and develop unique pathways of practice and social innovation. I am also optimistic that this book will add value to the larger field of action research, and it will further motivate many transdisciplinary professionals to engage with the same.

References

Chakraborty, I. (2018). *Stories at work: unlock the secret to business story telling*, Gurgaon Penguin Random House India.

Davenport, C., and Robertson, C. (2016). Resettling the first American 'climate refugees', *The New York Times*, 2 May 2016. Retrieved from www.nytimes.com/2016/05/03/us/resettling-the-first-american-climate-refugees.html [Last accessed on 2 September 2019].

Gandhi, M.K. (1961). *In search of the supreme*, compiled and edited by V.B. Kher, Vol. 1, Ahmedabad: Navajivan Publishing House.

ILO. (2019). *Working on a warmer planet: the impact of heat stress on labour productivity and decent work*, Geneva: International Labour Organisation.

Kumarappa, J.C. (1957). *Economy of permanence: a quest for social order based on non-violence*, 3rd edition, Varanasi: Sarva Seva Sangh Prakashan.

Rigaud, K.K., de Sherbinin, A., Jones, B., Bergmann, J., Clement, V., Ober, K., Schewe, J., Adamo, S., McCusker, B., Heuser, S., and Midgley, A. (2018). *Groundswell: preparing for internal climate migration*, Washington, DC: The World Bank.

Spretnak, C., and Capra, F. (1984). *Green politics*, London: Paladin Grafton Books.

Wall, T. (2019). 'This is a wake-up call': the villagers who could be Britain's first climate refugees, *The Guardian*, 18 May 2019. Retrieved from www.theguardian.com/environment/2019/may/18/this-is-a-wake-up-call-the-villagers-who-could-be-britains-first-climate-refugees [Last accessed on 20 October 2019].

2 Locating climate change adaptation

Introduction

The world is today witnessing climate change. There has been an unprecedented increase in the scale of human activity that has led to the disruption of diverse social–ecological systems. Global warming has induced rising of sea levels, severe storms, increased flooding and drought-like situations and scarcity of freshwater, which, in turn, have aggravated the vulnerabilities of millions of poor and marginalised populations across the globe. These events make it more difficult for vulnerable groups to secure their livelihoods and live with dignity. Be it in terms of variations in weather patterns or the frequent exposure of its population to the impacts of climate-related extreme events, nations and governments across the globe are gearing towards seeking effective adaptation measures. Recognition of localised, people-centred knowledge systems and decision-making are crucial in this context. In this regard we, as social workers, have a crucial role in strengthening resilient people-centred adaptation processes that are rooted in local knowledge systems and situated practices. Though our practice situations may not always have a direct mandate to address climate change, there is ample potential to integrate our values, ethics and ways of work with the goal of ensuring fair adaptation through our practices of care and solidarity. This book is about following one such adaptation pathway.

Adaptation in a systemic sense refers to those processes, actions or outcomes that could enable a system or its elements to better cope with, manage or adjust to some changing condition, stress, hazard, risk or opportunity (Smit and Wandel, 2006). With changing climatic conditions, people have to adjust to new forms of risks and uncertainties. The Intergovernmental Panel on Climate Change (IPCC) defines climate change adaptation as

> the process of adjustment to actual or expected climate and its effects. In human systems, adaptation seeks to moderate or avoid harm or exploit beneficial opportunities. In some natural systems, human intervention may facilitate adjustment to expected climate and its effects.
>
> (IPCC, 2014, p. 118)

Nevertheless, adaptation processes are not simple adjustments in human behaviour and are largely shaped by diverse contextual factors including dynamic power relations and situatedness of social actors. Adaptation has to be construed as socio-political processes that mediate how diverse actors deal with multiple and concurrent social–ecological changes (Eriksen et al., 2015). These processes are also multi-layered and dynamic in nature with discrete capacities, actions and outcomes (Pelling, 2011).

There are diverse standpoints to adaptation such as adjustment, reformist or transition and transformative approaches (Basett and Fogelman, 2013). The adjustment approach resembles top-down intervention strategies that are largely guided by techno-centric traditions, neglecting the socio-cultural and political nature of population vulnerability (Klepp and Chavez-Rodriguez, 2018). Such approaches aim at continuing with the status quo or functionality of the social–ecological system in a changing environment (Pelling, 2011). Adaptation projects with social transition goals aims at bringing about incremental changes in governance regimes and practices (Kronlid, 2014; Pelling, 2011). These projects maintain the essence and integrity of the existing technological, institutional, governance and value systems at a given scale (Agard and Schipper, 2014; Noble and Huq, 2014). To some extent, it reflects the system-based moral standpoints valuing authority, stability and preservation of social–ecological systems (Skoglund and Jensen, 2013). Transformative approaches in contrast look at adaptation and vulnerability as context-specific and ask for altering inherent social and economic contradictions, and therefore a more radical system change (Basett and Fogelman, 2013). They aim at bringing about pervasive changes in the fundamental attributes of the social–ecological system at a greater scale and institutional level, and strive towards altering discourses, values and power structures (Kronlid, 2014; Noble and Huq, 2014; Pelling, 2011).

The vulnerability-based moral standpoints on climate ethics give importance to solidarity, ability, need and entitlement, fairness in burdens and protection from harm (Adger et al., 2017). Such a standpoint critiques the approach of considering ecosystems passively, and instead advocates towards protection of marginalised and vulnerable groups as essential for justice, human capability and dignity (Draper and McKinnon, 2018). Yet another important standpoint is associated with a value-based approach to climate change adaptation. Values among diverse actors are necessarily not random constructions. Instead they are organised and integrated, coherent structures that are linked to the actors' motivations and actions (O'Brien and Wolf, 2010). The value-based approach emphasises on the aspect that actors maintain subjective and qualitative worldviews to climate change, which are important and unique to certain groups of actors and cultures. This perspective implies that neither there is a single objective way for adapting to climate change, nor there is a universal notion of climate justice that can be applied to all contexts. An understanding of these variations and interconnections in these standpoints are crucial in determining the processes and directions of our practice as well.

This book advocates an ethical standpoint of adaptation that is embedded in values of justice, care and solidarity. The adaptive innovation model proposed in this book has its foundation on certain fundamental constructs. The following sections of this chapter discusses these constructs in detail.

Community actors, situated knowledge and social–ecological systems

The term social–ecological system refers to an integrated understanding of humans-in-nature (Berkes et al., 2003). Such a perspective recognises that humans and ecosystems are intricately interconnected, each affecting the other and often in unpredictable ways (Cote and Nightingale, 2012). Climate change adaptation implies that we are dealing with complex social–ecological systems that involve a dynamic interface between diverse social actors and their natural environment (Aase et al., 2013; Berkes et al., 2003). In this regard, adaptive capacity of a social–ecological system could be understood as its ability to handle change and reconfigure itself without significant decline in crucial functions (Folke, 2006). Some of the key factors that could strengthen the adaptive capacities of actors in specific social–ecological systems are culturally robust local institutions, good leadership, effective cross-scale communication, integrated knowledge systems in resource management, political space for experimentation, equity in resource access and use of memory and knowledge as source of innovation and novelty (Santha, 2007; Seixas and Berkes, 2003; Sheridan and Nyamweru, 2008). A history of maintained connection between the people and their resource base are key to enhanced adaptive capacities (Tengö and von Heland, 2012). Local, decentralised institutions that facilitate collective action and innovation are found to be successful in adapting as they tend to be flexible and are able to respond more quickly to risk and uncertainty than centralised institutions (Colding et al., 2003). Aspects such as entitlements, livelihood security and access to natural resources are other crucial components that determine the adaptive capacities of the poor and marginalised populations (Sen, 1999).

S.F.F.2.1 Mangroves as a social–ecological system.

The Indian sub-continent is characterised by diverse unique social–ecological systems. In the last two decades, I had the opportunity to work with several traditional coastal communities in India. The drastic transitions happening to wetlands and coastal ecosystems have often raised my concern towards the sustainability of these social–ecological systems. Most of these ecosystems and people in these habitats are today vulnerable to extreme climatic events. For instance, let us take the case of mangrove-based social–ecological systems. Several traditional fishing households and indigenous communities along the tropical and sub-tropical coastlines depend on mangroves for their everyday livelihoods. These mangrove forests provide a wide range of ecosystem services such as fish, shellfish, timber and fuelwood. These forests also act as a strong buffer against major storms and tidal waves. They play a significant role in nurturing and protecting different varieties of marine organisms. Nevertheless, this unique social–ecological system is believed to disappear from the face of our earth at a rate of 0.66 per cent per year (FAO, 2007). Anthropogenic factors such as urbanisation, deforestation, encroachment, land reclamation, infrastructure development, dumping of waste, aquaculture and disruption of natural drains have significantly contributed to the decline of mangrove forests. In countries like India and Bangladesh, the destruction of mangrove forests has weakened the capacity of coastal and island communities to withstand storms and sea surges. Consequently, it has also affected the traditional livelihoods prevalent in the region. With the depletion of mangroves, adjoining farmlands in some locations have become extremely saline, and have made agriculture-based livelihoods such as paddy farming unviable. Consequently, many farming households are these days diversifying their livelihoods to shrimp-based aquaculture. However, the spread of aquaculture could further threaten the sustainability of the already fragile social–ecological system. More forests are being cleared for aquaculture without understanding the systemic interconnectedness. In this context, we also need to be aware of how certain livelihood diversification strategies of vulnerable groups could end up being maladaptive in the long run. On a positive note, I have come across several civil society organisations in partnership with local communities striving to restore these ecosystems. Organisations such as the M.S. Swaminathan Research Foundation (MSSRF) in partnership with local communities and state actors have developed innovative participatory co-management models to restore such fragile social–ecological systems.

Actors constitute individuals, groups and organisations within a specific social–ecological system. In a specific social–ecological system, community actors consist of individuals, groups and organisations having certain shared values, commitments, ideologies, knowledge and interests. The term 'community' is quite ambiguous and could be applied differently at the same time by diverse actors to justify different politics, policies and practices (Mayo, 2008; Mulligan, 2018). We should be aware that the notion of community does not represent a homogenous social structure. Instead, it has to be conceived as 'localised, fragmented, hybrid, multiple, overlapping and activated differently in different arenas and practices' (Rose, 1999, p. 178). For many vulnerable communities, their local knowledge is the sole enabler to decide and devise a suitable adaptation strategy (Bankoff et al., 2012). Recognition of local knowledge systems is therefore crucial towards designing people-centred adaptation strategies (Kendrick, 2003).

Local knowledge refers to the awareness, understanding and skills of community actors, which they gain through their everyday interaction and experiences within specific social–ecological systems. It evolves over time, which also provides local communities the capacity to adapt to risks, uncertainties and related insecurities (Dixon, 2005; Riggs, 2005; Sillitoe, 1998). Local knowledge systems guarantee a sense of freedom, trust and autonomy for its custodians (Sillitoe, 2010). Community actors as custodians of local knowledge also engage in a form of science along with their everyday struggles of survival and subsistence (Barnhardt, 2005; Jigyasu, 2002; Parvin et al., 2009). It further entails the reproduction of the community and their expressions to the core values entwined with the social (Appadurai, 1990a). In this context, memory as part of accumulated experience and history of the system is an important component of community actors' local knowledge. Rituals and practices associated with these memories provide ample scope for the social–ecological system to reorganise after a disturbance or be prepared, anticipating future disturbances (Berkes et al., 2003). It also enables vulnerable communities to question the conventional views of adaptation and development.

Local knowledge systems are closely interlinked to other components of culture in a specific social–ecological system. For instance, studies have shown that local knowledge supplements prevailing values and norms in the sustainable management of common property resources such as forests and fisheries (Santha, 2008). These interlinkages also help in avoiding resource-use conflicts among many natural resource-dependent communities. However, culture, local knowledge and the institutions mediating their interlinkages with the natural environment are not static and will undergo changes due to the influence of various contextual factors. Various factors such as changes in family structure, intervention by state and market actors, geographical specificity, interactions with other forms of expert knowledge and resource scarcity could alter the way humans interact with their natural environment (ibid). Changes in the physical environment, economy, social structures and cultural practices themselves could become sources of uncertainty while adapting to climate change. It is therefore crucial to understand this dynamic nature of culture and local knowledge while designing institutions for adaptation in specific social–ecological systems. Nevertheless, it will be challenging for us to understand the dynamics of social–ecological systems at varying scales and there could be knowledge barriers and uncertainties even at the grassroots level (Berkes et al., 2003; Rizvi and van Riel, 2014). Whatever the challenges may be, we have to strive towards the co-creation of bottom-up, people-centred adaptation designs by collaborating with community actors and other relevant stakeholders.

S.F.F.2.2 Role of local knowledge in climate change adaptation.

There are numerous traditional communities who have developed their adaptation strategies based on their local knowledge. They have their own ways of forecasting floods, drought and other extreme weather events. For instance, coastal fishing communities in South India rely on the pattern and directions of wind to forecast storms and sea surge. According to them, 'the wind from south-west direction is a sign of severe storm'. On the other hand, they believe that 'if the wind begins to blow from the north to east, the sea will become calmer'. However, they also note that 'if the wind blows from the south and the oceanic currents are from the north, it would result in severe surge and coastal erosion'. Howell (2003) has mentioned about similar knowledge systems prevalent among coastal communities in Bangladesh as well. In a similar vein, certain tribal communities anticipate drought-like situations if certain varieties of mushrooms grow in abundance before the rainy season (Figaredo, 2009). I have observed that farmers rely on a kind of practical knowledge to experiment new ideas and imaginations in their everyday lives. For example, farmers tend to cultivate crops like legumes to restore soil and crop fertility in their fields. I have met several marginal farmers who sustain their livelihood and ecological security through mixed cropping strategies. According to them, such integrated farm management strategies prevent the main crop from being affected by pests, wind or other hazards. In a similar vein, I have read that farmers in Kenya have piloted a push–pull system to resist pest attack and improve their livelihood security (Kassie et al., 2018; Pretty and Bharucha, 2015).

Knowledge production in community settings also have to be seen as a process of social negotiation involving multiple actors and complex power relations (Clarke, 2005; Nygren, 1999). Community actors attempt to make sense of an emerging situation by relating to what they already know about that situation (Hathcoat and Nicholas, 2014; Lang, 2011; Leino and Peltomaa, 2012). Each actor would accord meanings and construct their respective situations differently by bringing their unique personal experiences, commitments, social locations and subjective creativity to the process of assembling knowledge (Code, 1991; Lang, 2011; Sunil, 2019). Moreover, these meanings are not fixed, as the so constructed knowledge is constantly open to change and modification (Code, 1982). One's social situation both enables and sets limits on what one can know (Code, 2006; Rosendahl et al., 2015). As practitioners, we need to be aware of how community actors strategically use their own and others' knowledge to fulfil their interests and how such practices result in certain context-specific outcomes (Long, 2001; Srang-iam, 2013). In this regard, we have to give attention to the politics of difference in the production of knowledge (Hartman, 1994; Richardson-Ngwenya, 2013). The everyday social encounters reflect how diverse actors would manoeuvre and exercise power in specific situations (Long and Long, 1992). Each social encounter in climate change adaptation could involve diverse community actors outlining their identities, affiliations and discourses pertaining to their past and present circumstances, as well as in claiming legitimacy for their social and political action (Eriksen et al., 2015; Schetter, 2005). These interactions are usually routinised across intersectional dimensions of class, gender, race, caste, ethnicity, region, religion, language or occupation and their natural environment. In these contexts, conceptualising knowledge production as merely local and in functionalist terms may limit us from recognising the plural, partial and intersectional nature of knowledge. A deeper engagement is possible through the lens of situated knowledge.

Situated knowledge refers to the local, specific knowledge that has its value on the particular situation at hand (Haraway, 1988). It is critical in nurturing shared conversations and sustaining solidarity networks (Haraway, 1988; Leino and Peltomaa, 2012). It recognises the collective historical subjectivity and agency of embodied actors (Haraway, 1988). The focus of adaptive innovation is on situational and practical form of knowing that is generated through participative and collaborative interaction. Our role as social workers is to recognise the dynamic and emergent nature of social encounters and evolve innovative and appropriate methodologies of inquiry and practice to work with the situatedness of actors' specific knowledge, social positions and lived experiences. On many instances, community actors collaborate and settle with other external and institutional actors for a mutually agreed-upon knowledge (Caretta, 2015; Kobayashi, 1994; Nightingale, 2003). It therefore becomes important to examine each social encounter in terms of how certain forms of knowledge interfaces sustain or reproduce inequalities (Abraham and Purkayastha, 2012). In this regard, several studies have shown how the knowledge and experiences of marginalised actors and vulnerable groups are crucial to understand the implications of development interventions (Berkes, 1999; Houston, 2010; Marglin and Marglin, 1990; Richardson-Ngwenya, 2013; Rosendahl et al., 2015; Scott, 1998).

We have to be attentive to the contexts of engagement in which the knowledge of every actor is applied (Lauer and Aswani, 2009). In this regard, community actors who are dependent on natural resources should be enabled to innovate through the nurturing, restoration and conservation of their resource base; while whose livelihoods are no longer dependent on them should be provided adequate learning contexts to reflect and reconnect to their land, water or forests (Davidson-Hunt and Berkes, 2003). Local community actors who are closely connected to a resource system will be in a better position to adapt, as they could be motivated to develop practices and structures that may reduce the effect of complex environmental changes (Carlsson, 2003; Colding et al., 2003). Focus should be on the agency and multiple standpoints of the subjugated actors. More specifically, we have to consistently explore and be alert to the experiences of those community actors that could remain unexamined (Abraham and Purkayastha, 2012). Locating oneself in such situated practices could help in recognising and accommodating the hybridity and heterogeneousness of knowledge as well (Ellen, 2004; Lauer and Aswani, 2009; Nygren, 1999; Srang-iam, 2013). We have to therefore value the significance of shared conversations and dialogic interactions as a preferred methodology in our situations of practice (Haraway, 1988; Stoetzler and Yuval-Davis, 2002; Tickner, 1997). In this context, our role as social workers is to explore ways for co-creating shared spaces, interests and destiny that would recognise the historical experiences and future aspirations of community actors (Jha, 2010). Opening up to other narratives and knowledge systems could show new possibilities of representing subjugated knowledge (Hartman, 1994).

S.F.F.2.3 Deep-water rice cultivation as an innovative adaptation.

Some of my students were engaged with the livelihood enhancement of Musahar communities in Bihar, India. These communities are one of the most historically marginalised and socially excluded groups in India. Since the last few decades, they have been victims of severe riverbank erosion. Most of these communities are landless. And those people who have managed to acquire some land for cultivation are in the fear of losing them to riverbank erosion. Though these communities are widespread across Bihar, I have come to know about an innovative farming practice that a particular Musahar community follows. This community is located in between the embankments of two rivers, namely Kosi and Kamla Balan in North Bihar. Since many decades, these communities have been involved in the farming of a deep-water rice variety called the '*Desaria Dhan*'. This particular paddy is flood-resistant and is a floating variety of deep-water rice. It has the ability to grow five–six feet and float in deep floodwater (Catling, 1992). However, such unique practices of adaptation are facing severe threat due to higher levels of flooding caused from infrastructure development along the rivers (Jamwal, 2017). Constructions of hard structures such as sluice gates and concrete river embankments without recognising the indigenous practices of these communities have placed their livelihoods at stake. In this regard, it is utmost important that any attempts towards developing climate change adaptation strategies should empower local communities and support local innovation rather than impose external solutions.

Ideas, imagination and practice

Ghosh (2016) envisages the climate crisis as a moral and cultural predicament, and therefore as a crisis of imagination. Imagination is a social practice (Appadurai, 1990b). In a sense, it represents the narrative side of practice (Bakonyi, 2015). In their everyday social encounters, different community actors are involved in specific practices of framing distinct narratives on their situation and developing ideas about what could be done. Imagination gives community actors a direction between what 'is' and what 'ought to be' (Stoetzler and Yuval-Davis, 2002). It is also an aspirational pathway for diverse actors to construct and engage in rational conversations and meaningful dialogue (Haraway, 1991; Stoetzler and Yuval-Davis, 2002). Like knowledge, imagination is also situated, shaped and conditioned by the social position and historical location of actors (Appadurai, 1990b; Stoetzler and Yuval-Davis, 2002). It gives specific meaning to their experiences and has the potential to transcend pre-fixed boundaries and alter existing social realities (Stoetzler and Yuval-Davis, 2002).

Imagination is a pre-requisite for the co-creation of actionable knowledge. Ideas shape politics, decision-making processes and the framing of action (Bakonyi, 2015). It further adds value to the agency of community actors to bring about desired social change (Stoetzler and Yuval-Davis, 2002). Each imagery of social change consists of a chain of ideas, terms and images that ultimately would help community actors to claim authority and power (Appadurai, 1990b). These ideas and imageries emerge, disappear or re-emerge during specific social encounters. It is equally important to understand and analyse how these ideas and imageries are interpreted, manifested and enacted by community actors in different socio-historical and spatial contexts (Martinez, 2012). Ideas, when adapted to local conditions, interact with locally predominant narratives. These narratives could then become part of the actors' imagination and shape the way in which they make sense of their own social realities (ibid). While there will be many community actors imagining change, there can be some actors who are more powerful in defining and controlling these imaginations. In this regard, we need to empathetically recognise the experience, imagination and knowledge of marginalised actors (Stoetzler and Yuval-Davis, 2002).

At the same time, we need to be alert to our own imaginations on adaptation and development, where we could end up viewing recipient communities as beneficiaries of outside expert interventions in contrast to their self-determination and inherent capacities. We could also transpire being one among the prominent actors attempting to steadily reinforce our own ideas and impressions on local community actors. There is a strong need therefore to be alert to and regularly self-reflect on our own positionalities, ideas, imaginations and practice. As mentioned earlier, the emphasis has to be towards the co-creation of knowledge and action through shared conversations and dialogue among different actors. Such collaborative dialogic processes could result in genuine innovation and emergence of new practices and institutions (Hajer and Wagenaar, 2003). Such a practice-based view would enable us to co-create climate change adaptation as a contextual, process-oriented and actor-networked approach.

Practice situations involve actors, actants, knowledge, discourse, structure/process and agency and, importantly, these elements cannot be considered in isolation (Clarke, 2005). The practices of community actors are shaped by a set of activities, routines, material artefacts, memories, imaginations and diverse negotiation strategies embedded within each social encounter (Schatzki, 2005; Wagenaar and Cook, 2003). A series of routine activities alone cannot be declared as practice, as other social and cultural components such as performances and representations or talk that are characteristic of each social encounter could also constitute practice (Schau et al., 2009; Wagenaar and Cook, 2003). Such a perspective on practice also highlights the interactive, context-bound character of actions (Wagenaar and Cook, 2003). Each social encounter has to be looked upon as a context where imaginations and counter-imaginations of various actors in diverse realms of power strive towards retaining or acquiring power through practice. Community actors in practice situations are never a detached observer. Imaginations of one's own self and how he or she wants others to see him or her are crucial factors that shape practice (ibid). It implies that in negotiating a particular situation, the actor is always aware of his or her position in a larger network of relations and obligations (Hajer and Wagenaar, 2003). Practice is thus shaped by one's social position as well as the contexts in which each actor listen, negotiate and participate in the situatedness of other fellow actors (Wagenaar and Cook, 2003).

Action, the core element of any conception of practice is also a way of knowing, which accounts for both the actor and the situation. It enables people to negotiate the world by acting upon it and creating new forms of knowledge and practice (Seur, 1992; Wagenaar and Cook, 2003). It is through action that community actors recreate or alter their practice situations (Verschoor, 1992). The strategic nature of actions and their outcomes have to be examined through reflective practice. Action accompanied by reflection could not only produce change within the surroundings, but also lead to self-awareness and lend voice to community actors' interests that matter the most (Seur, 1992). It provides each actor with further clarity on the situation and therefore on what might be possible (ibid). In this regard, action research approaches have proven to be a suitable methodology.

Action research and innovation platforms

The adaptive innovation model discussed in this book has largely drawn insights from the action research paradigm. Action research, in this book is referred to as a transformative change-making approach guided by the values of justice, care and solidarity that are constituted through participatory inquiry and reflective practice. Action research is a methodology that is broadly defined and takes widely different forms (Somekh, 1995). Reason and Bradbury (2008) view action research as more of an orientation to inquiry, which seeks to create participative communities focused on action, reflection and practice in the pursuit of practical solutions to issues of concern. It is a change-mediating process that has the potential to transform people's lives by their own collective efforts (Stark, 2014). In this regard, action research is a set of intuitive and reflective processes that are highly context specific and not necessarily designed by any prior categorisation of knowledge-constitutive interests (Elliott, 2005). A key purpose of action research is 'to forge a more direct link between intellectual knowledge and moment-to-moment personal and social action, so that inquiry contributes directly to the flourishing of human persona, their communities and the ecosystems of which they are part' (Reason and Torbert, 2001, p. 6).

Action research frameworks have their core values in community participation, critical self-awareness and social justice (George and Syrja-McNally, 2015; Reitan and Gibson, 2012). These frameworks are also consistent with that of other social work values, especially those pertaining to partnership, relationship and social change (Barbera, 2008; Katwyk and Ashcroft, 2016). Action research values are largely shaped by the participatory paradigm, which considers relationships as pivotal to the validity of both knowledge and the process of knowledge production (Lincoln et al., 2011; Loewenson et al., 2014). It strives towards addressing issues related to positions of privilege and power; and aims at empowering participant actors to take control of the dominant factors shaping their lives (Fals-Borda, 1987; Healy, 2001; Tandon, 1981). It encourages participant actors to be empathetic towards marginalised perspectives and subjugated knowledges among them.

Climate change adaptation facilitated through action research will not only provide a new understanding to the phenomena and practice of adaptation, but also has the scope to blend understanding with change or action. There is a significant body of literature advocating the merits of action research in climate change adaptation and sustainable transitions (Amaru and Chhetri, 2013; Bradbury et al., 2019; Campos et al., 2016; Morchain et al., 2019). Studies have demonstrated how action research has enabled farmers to enhance their adaptive capacities in terms of their willingness to experiment, trust their own experimentation and informal networks, cross-fertilise expert knowledge, take strategic decisions, introduce new technology, resolve conflicts and mitigate risks (Arbuckle et al., 2015; Bloch et al., 2016; Kerr et al., 2018; Mapfumo et al., 2013). On certain occasions, the iterative and reflective approach that action research facilitates has enabled organic farmers and agricultural scientists to collaborate and co-create innovative adaptation strategies (Bloch et al., 2016). In a similar vein, action research projects have enabled co-participants in climate justice endeavours to explore diverse adaptation options to resolve complex issues that otherwise did not pose obvious solutions (Hall et al., 2009; Tanner and Seballos, 2012). Action research embedded with reflective practice also has the potential to enhance the agency or action competence of co-participants in climate justice initiatives (Hall et al., 2009). Through the medium of action research, we will also be capable of understanding the discontinuities associated with collective groups and specifically on the nature of frictions, disagreements and conflicts between community actors, which are mediated and transformed at critical junctures (Santha et al., 2017).

Rooted in action research, the adaptive innovation model envisages the nurturing of ethical climate adaptation strategies through community-based innovation platforms. Innovation platforms in the context of climate change adaptation could be understood as a collective and collaborative institutional space owned and governed by community actors in partnership with other relevant stakeholders to mutually share knowledge, imagine and innovate situated practices that could reduce vulnerabilities and build resilience of social–ecological systems. Certain defining elements of a community-based innovation platform could be listed as follows,

- Its intimate connectedness to the primary livelihoods and social–ecological system of community actors;
- The knowledge of participant actors pertaining to the dynamics of social–ecological system and on the felt impact of these changes;

- A shared commitment and willingness among participant actors to imagine, ideate and test diverse adaptation options;
- Flexible, but mutually agreed-upon rules for appropriation and provision from the social–ecological system; and
- Nested institutional networks and resources to facilitate the scaling up of piloted solutions.

Innovation platforms have ample scope to emerge as spaces where everyday learning, reflection, dialogue and experimentation by community actors are routinized. The structural and interactional characteristics of innovation platforms would shape the nature of collective action among actors. The flexible, but unique and distinct set of norms, values, beliefs, behaviours, styles, mode of communication and even language of participant actors shape the identity and ethos of the innovation platform. Values and practices that nurture inclusive, democratic and constructive involvement of actors are likely to result in a greater interest and participation of community actors in the innovation platform. A sustained effort towards adapting to climate change demands an active engagement of participant actors in a continuous process of critical reflection and action across diverse social–ecological systems (Amaru and Chhetri, 2013). Such a strategy will also provide ample opportunities for us to build new perspectives on ethical adaptation and develop alternate set of skills to decide and act in domains that are uncertain and complex (Gould and Taylor, 1996; Gray and Coates, 2015; Joseph, 2017).

S.F.F.2.4 Village-level institutions as innovation platforms.

There are different types of village-level institutions that can serve as effective innovation platforms. These can be cooperatives, self-help groups or Village Development Committees (VDCs). What matters most is how members of the platform understand their contexts, identify their needs, prioritise their actions, form rules and regulations, implement and monitor them. There are several examples in South Asia, where village-level institutions have enabled communities to overcome ecological crisis and improve their well-being and quality of life. In this regard, I would comment that the VDC of Lapodiya village in Rajasthan, India could be considered as an active innovation platform. The VDC has good leadership, community collaboration and is self-driven by their knowledge systems and passion for innovation. They have evolved context-specific natural resource management norms and sanctions embedded in the values of care and solidarity. The VDC has prohibited the cutting of trees and any violation of this norm is met with a sanction of planting additional trees. In a similar vein, birds are not supposed to be harmed, as they are considered as agents of faster afforestation. Each house has arrangements to feed water and food to the birds. A grain bank has also been established to ensure both food and seed security. This village was once a severely water scarce village highly marked by deprivation, poverty and migration. Today, experts observe that this village has achieved self-sufficiency in water management by restoring traditional water management systems as well as through inventing unique water harvesting systems called the '*chauka*' (Padre, 2017).

This chapter has given a brief overview of the diverse concepts and contexts that would help us to comprehend the model of adaptive innovation. Having discussed the key concepts that could guide the process of adaptive innovation, the next chapter elaborates the landscape of adaptive innovation in detail.

References

Aase, T.H., Chapagain, P.S., and Tiwari, P.C. (2013). Innovation as an expression of adaptive capacity to change in Himalayan farming, *Mountain Research and Development*, 33, pp. 4–10.

Abraham, M., and Purkayastha, B. (2012). Making a difference – linking research and action in practice, pedagogy, and policy for social justice: introduction, *Current Sociology*, 60(2), pp. 123–141.

Adger, W.N., Butler, C., and Walker-Springett, K. (2017). Moral reasoning in adaptation to climate change, *Environmental Politics*, 26(3), pp. 371–390.

Agard, J., and Schipper, L. (Eds.). (2014). Glossary, in *Climate Change 2014: Impacts, Adaptation, and Vulnerability. Working Group II Contribution to the IPCC 5th Assessment Report*, Geneva, Switzerland: IPCC, pp. 1–30.

Amaru, S., and Chhetri, N.B. (2013). Climate adaptation: institutional response to environmental constraints, and the need for increased flexibility, participation, and integration of approaches, *Applied Geography*, 39, pp. 128–139.

Appadurai, A. (1990a). Technology and the reproduction of values in rural western India, in F.A. Marglin and S.A. Marglin (Eds.). *Dominating knowledge: development, culture and resistance*, Oxford: Clarendon Press, pp. 185–216.

Appadurai, A. (1990b). Disjuncture and difference in the global cultural economy, in J.X. Inda and R. Rosaldo (Eds.). (2002). *The anthropology of globalization: a reader*, Oxford: Blackwell Publishing, pp. 46–64.

Arbuckle, J.G., Jr., Morton, L.W., and Hobbs, J. (2015). Understanding farmer perspectives on climate change adaptation and mitigation: the roles of trust in sources of climate information, climate change beliefs, and perceived risk, *Environment and Behavior*, 47(2), pp. 205–234.

Bakonyi, J. (2015). Ideoscapes in the world society: framing violence in Somalia, *Civil Wars*, 17(2), pp. 242–265.

Bankoff, G., Frerks, G., and Hilhorst, D. (2012). *Mapping vulnerability: disasters, development and people*, London: Routledge.

Barbera, R.A. (2008). Relationships and the research process: participatory action research and social work, *Journal of Progressive Human Services*, 19(2), pp. 140–159.

Barnhardt, R. (2005). Indigenous knowledge systems and Alaska native ways of knowing, *Anthropology and Education Quarterly*, 36, pp. 8–23.

Basett, T.J., and Fogelman, C. (2013). Dejavu or something new? the adaptation concept in climate change literature, *Geoforum*, 48, pp. 42–53.

Berkes, F. (1999). *Sacred ecology: traditional ecological knowledge and resource management*, Philadelphia, PA: Taylor and Francis.

Berkes, F., Colding, J., and Folke, C. (2003). *Navigating social-ecological systems: building resilience for complexity and change*, Cambridge: Cambridge University Press.

Bloch, R., Knierim, A., Häring, A., and Bachinger, J. (2016). Increasing the adaptive capacity of organic farming systems in the face of climate change using action research methods, *Organic Agriculture*, 6, pp. 139–151.

Bradbury, H., Waddell, S., O'Brien, K., Apgar, M., Teehankee, B., and Fazey, I. (2019). A call to action research for transformations: the times demand it, *Action Research*, 17(1), pp. 3–10.

Campos, I.S., Alves, F.M., Dinis, J., Truninger, M., Vizinho, A., and Penha-Lopes, G. (2016). Climate adaptation, transitions, and socially innovative action-research approaches, *Ecology and Society*, 21(1), p. 13. Retrieved from http://dx.doi.org/10.5751/ES-08059-210113 [Last accessed on 10 November 2019].

Caretta, M.A. (2015). Situated knowledge in cross-cultural, cross-language research: a collaborative reflexive analysis of researcher, assistant and participant subjectivities, *Qualitative Research*, 15(4), pp. 489–505.

Carlsson, L. (2003). The strategy of the commons: history and property rights in central Sweden, in F. Berkes, J. Colding, and C. Folke (Eds.). *Navigating social-ecological systems: building resilience for complexity and change*, Cambridge: Cambridge University Press, pp. 116–131.

Catling, D. (1992). *Rice in deep water*, London: The Macmillan Press Limited and International Rice Research Institute.

Clarke, A.E. (2005). *Situational analysis: grounded theory after the postmodern turn*, Thousand Oaks, CA: Sage.

Code, L. (1982). The importance of historicism for a theory of knowledge, *International Philosophical Quarterly*, 22(2), pp. 157–174.

Code, L. (1991). *What can she know? Feminist theory and the construction of knowledge*, Ithaca, NY: Cornell University Press.

Code, L. (2006). *Ecological thinking: the politics of epistemic location*, Cheshire Calhoun, Oxford: Oxford University Press, pp. 40–41.

Colding, J., Elmqvist, T., and Olsson, P. (2003). Living with disturbance: building resilience in social–ecological systems, in F. Berkes, J. Colding, and C. Folke (Eds.). *Navigating social-ecological systems: building resilience for complexity and change*, Cambridge: Cambridge University Press, pp. 163–185.

Cote, M., and Nightingale, A.J. (2012). Resilience thinking meets social theory: situating social change in social-ecological systems (SES) research, *Progress in Human Geography*, 36(4), pp. 475–489.

Davidson-Hunt, I.J., and Berkes, F. (2003). Nature and society through the lens of resilience: toward a human-in-ecosystem perspective, in F. Berkes, J. Colding, and C. Folke (Eds.). *Navigating social-ecological systems: building resilience for complexity and change*, Cambridge: Cambridge University Press, pp. 53–82.

Dixon, A. (2005). Wetland sustainability and the evolution of indigenous knowledge in Ethiopia, *The Geographical Journal*, 171, pp. 306–323.

Draper, J., and McKinnon, C. (2018). The ethics of climate-induced community displacement and resettlement, *WIRE's Climate Change*, 9, p. e519.

Ellen, R. (2004). From ethno-science to science, or what the indigenous knowledge debate tells us about how scientists define their projects, *Journal of Cognition and Culture*, 4(3), pp. 409–450.

Elliott, J. (2005). Becoming critical: the failure to connect, *Educational Action Research*, 13(3), pp. 359–373.

Eriksen, S.H., Nightingale, A.J., and Eakin, H. (2015). Reframing adaptation: the political nature of climate change adaptation, *Global Environmental Change*, 35, pp. 523–533.

Fals-Borda, O. (1987). The application of participatory action research in Latin America, *International Sociology*, 2(4), pp. 329–347.

FAO. (2007). *The world's mangroves 1980–2005*, Forestry Paper No. 153, Rome: Food and Agriculture Organisation of the United Nations. Retrieved from www.fao.org/3/a1427e/a1427e04.pdf [Last accessed on 23 October 2019].

Figaredo, L.B. (2009). Tribal communities and adaptation to climate change, Pipal Tree – Fireflies, Paper presented at Conference on 'What is the good life? – Exploring faith-based and secular values and action-perspectives to mitigate climate change', 8–11 October 2009; *Pipal Tree*, Fireflies Inter-Cultural Centre, Bangalore.

Folke, C. (2006). Resilience: the emergence of a perspective for social–ecological systems analyses, *Global Environmental Change*, 16, pp. 253–267.

George, P., and Syrja-McNally, D. (2015). Social enquiry and action research for social work, in J.D. Wright (Ed.). *International encyclopedia of the social and behavioural sciences*, 2nd edition, Vol. 22, Oxford: Elsevier, pp. 269–274.

Ghosh, A. (2016). *The great derangement: climate change and the unthinkable*, Gurgaon: Penguin and Allen Lane.

Gould, N., and Taylor, I. (Eds.). (1996). *Reflective learning for social work*, Aldershot: Arena.

Gray, M., and Coates, J. (2015). Changing gears: shifting to an environmental perspective in social work education, *Social Work Education*, 34(5), pp. 502–512.

Hajer, M., and Wagenaar, H. (2003). Introduction, in M. Hajer and H. Wagenaar (Eds.). *Deliberative policy analysis: understanding governance in the network society*, Cambridge: Cambridge University Press, pp. 1–30.

Hall, N.L., Taplin, R., and Goldstein, W. (2009). Empowerment of individuals and realization of community agency: applying action research to climate change responses in Australia, *Action Research*, 8(1), pp. 71–91.

Haraway, D. (1988). Situated knowledges: the science question in feminism and the privilege of partial perspective, *Feminist Studies*, 14(3), pp. 575–599.

Haraway, D.J. (1991). *Simians, cyborgs and women: the reinvention of nature*, New York: Routledge.

Hartman, A. (1994). *Reflection and controversy: essays on social work*, Washington, DC: NASW Press.

Hathcoat, J.D., and Nicholas, M.C.N. (2014). Epistemology, in D. Coghlan and M. Brydon-Miller (Eds.). *The Sage encyclopedia of action research*, New Delhi: Sage, pp. 302–306.

Healy, K. (2001). Participatory action research and social work: a critical appraisal, *International Social Work*, 44(1), pp. 93–105.

Houston, S. (2010). Prising open the black box: critical realism, action research and social work, *Qualitative Social Work*, 9(1), pp. 73–91.

Howell, P. (2003). Indigenous early warning indicators of cyclones: potential application in coastal Bangladesh, Working Paper 6, University College London: Benfield Hazard Research Centre.

IPCC. (2014). Annex II: glossary, in K.J. Mach, S. Planton, and C. von Stechow (Eds.). *Climate Change 2014: Synthesis Report. Contribution of Working Groups I, II and III to the Fifth Assessment Report of the Intergovernmental Panel on Climate Change* [Core Writing Team, R.K. Pachauri and L.A. Meyer (Eds.)], Geneva, Switzerland: IPCC, pp. 117–130.

Jamwal, N. (2017). Flood resistant rice resists for survival, *India Climate Dialogue*, 18 October. Retrieved from https://indiaclimatedialogue.net/2017/10/18/flood-resistant-desariya-rice-fights-survival/ [Last accessed on 15 November 2019].

Jha, M. (2010). Community organisation in split societies, *Community Development Journal*, 44(3), pp. 305–319.

Jigyasu, R. (2002). *Reducing disaster vulnerability through local knowledge and capacity*, Doctorate of Engineering thesis, Department of Urban Design and Planning, Norwegian University of Science and Technology, Trondheim.

Joseph, D.D. (2017). Social work models for climate adaptation: the case of small islands in the Caribbean, *Regional Environmental Change*, 17, pp. 1118–1126.

Kassie, M., Stage, J., Diiro, G., Muriithi, B., Muricho, G., Ledermann, S.T., Pittchar, J., Midega, C., and Khan, Z. (2018). Push–pull farming system in Kenya: implications for economic and social welfare, *Land Use Policy*, 77, pp. 186–198.

Katwyk, T.V., and Ashcroft, R. (2016). Using participatory action research to access social work voices: acknowledging the fit, *Journal of Progressive Human Services*, 27(3), pp. 191–204.

Kendrick, A. (2003). Caribou co-management in northern Canada: fostering multiple ways of knowing, in F. Berkes, J. Colding, and C. Folke (Eds.). *Navigating social-ecological systems: building resilience for complexity and change*, Cambridge: Cambridge University Press, pp. 241–267.

Kerr, B.R., Nyantakyi-Frimpong, H., Dakishoni, L., Lupafya, E., Shumba, L., Luginaah, I., and Snapp, S.S. (2018). Knowledge politics in participatory climate change adaptation research on agroecology in Malawi, *Renewable Agriculture and Food Systems*, 33, pp. 238–251.

Klepp, S., and Chavez-Rodriguez, L. (2018). Governing climate change: the power of adaptation discourses, policies, and practices, in S. Klepp and L. Chavez-Rodriguez (Eds.). *A critical approach to climate change adaptation: discourses, policies and practices*, London: Routledge-Earthscan, pp. 3–34.

Kobayashi, A. (1994). Coloring the field: gender, "race", and the politics of fieldwork, *Professional Geographer*, 46, pp. 73–80.

Kronlid, D.O. (2014). *Climate change adaptation and human capabilities: justice and ethics in research and policies*, New York: Palgrave Macmillan.

Lang, C. (2011). Epistemologies of situated knowledges: "troubling" knowledge in philosophy of education, *Educational Theory*, 61(1), pp. 75–96.

Lauer, M., and Aswani, S. (2009). Indigenous ecological knowledge as situated practices: understanding fishers' knowledge in the Western Solomon Islands, *American Anthropologist*, 111(3), pp. 317–329.

Leino, H., and Peltomaa, J. (2012). Situated knowledge–situated legitimacy: consequences of citizen participation in local environmental governance, *Policy and Society*, 31(2), pp. 159–168.

Lincoln, Y.S., Lynham, S.A., and Guba, E.G. (2011). Paradigmatic controversies, contradictions, and emerging confluences, revisited, in N.K. Denzin and Y.S. Lincoln (Eds.). *The Sage handbook of qualitative research*, 4th edition, Thousand Oaks, CA: Sage, pp. 97–128.

Loewenson, R., Laurell, A.C., Hogstedt, C., D'Ambruoso, L., and Shroff, Z. (2014). Participatory action research in health systems: a methods reader. Retrieved from www.equi netafrica.org/sites/default/files/uploads/documents/PAR%20Methods%20Reader2014% 20for%20web.pdf [Last accessed on 12 September 2019].

Long, N. (2001). *Development sociology: actor perspectives*, London: Routledge.

Long, N., and Long, A. (Eds.). (1992). *Battlefields of knowledge: the interlocking of theory and practice in social research and development*, London: Routledge.

Mapfumo, P., Adjei-Nsiah, S., Mtambanengwe, F., Chikowo, R., and Giller, K.E. (2013). Participatory action research (PAR) as an entry point for supporting climate change adaptation by smallholder farmers in Africa, *Environmental Development*, 5, pp. 6–22.

Marglin, F.A., and Marglin, S.A. (Eds.). (1990). *Dominating knowledge: development, culture, and resistance*, Oxford: Clarendon Press.

Martinez, E.J. (2012). *On making sense: queer race narratives of intelligibility*, Stanford, CA: Stanford University Press.

Mayo, M. (2008). Community development: contestations, continuities and change, in G. Craig, K. Popple, and M. Shaw (Eds.). *Community development in theory and practice: an international reader*, Nottingham: Spokesman Books, pp. 13–27.

Morchain, D., Spear, D., Ziervogel, G., Masundire, H., Angula, M.N., Davies, J., Molefe, C., and Hegga, S. (2019). Building transformative capacity in southern Africa: surfacing knowledge and challenging structures through participatory vulnerability and risk assessments, *Action Research*, 17(1), pp. 19–41.

Mulligan, M. (2018). Rethinking the meaning of 'community' in community-based disaster management, in G. Marsh, I. Ahmed, M. Mulligan, J. Donovan, and S. Barton (Eds.). *Community engagement in post-disaster recovery*, London: Routledge, pp. 1–10.

Nightingale, A. (2003). A feminist in the forest: situated knowledges and mixing methods in natural resource management, *ACME: An International E-Journal for Critical Geographies*, 2(1), pp. 77–90.

Noble, I.R., and Huq, S. (2014). Adaptation needs and options, in *Climate Change 2014: Impacts, Adaptation, and Vulnerability*. Working Group II Contribution to the IPCC 5th Assessment Report, Final draft, Geneva, Switzerland: IPCC, pp. 1–51.

Nygren, A. (1999). Local knowledge in the environment–development discourse: from dichotomies to situated knowledges, *Critique of Anthropology*, 19(3), pp. 267–288.

O'Brien, K.L., and Wolf, J. (2010). A values-based approach to vulnerability and adaptation to climate change, *WIREs Climate Change*, 1, pp. 232–242.

Padre, S. (2017). *The water catchers*, New Delhi: Nimby Books.

Parvin, G.A., Takahashi, F., and Shaw, R. (2009). Coastal hazards and community coping methods in Bangladesh, *Journal of Coastal Conservation*, 12, pp. 181–193.

Pelling, M. (2011). *Adaptation to climate change: from resilience to transformation*, London: Routledge.

Pretty, J., and Bharucha, Z.P. (2015). Integrated pest management for sustainable intensification of agriculture in Asia and Africa, *Insects*, 6(1), pp. 152–182.

Reason, P., and Bradbury, H. (2008). Introduction to groundings, in P. Reason and H. Bradbury (Eds.). *The Sage handbook of action research: participative inquiry and practice*, 2nd edition, New Delhi: Sage, pp. 1–10.

Reason, P., and Torbert, W. (2001). The action turn: towards a transformational social science, *Concepts and Transformations*, 6(1), pp. 1–37.

Reitan, R., and Gibson, S. (2012). Climate change or social change? Environmental and leftist praxis and participatory action research, *Globalizations*, 9(3), pp. 395–410.

Richardson-Ngwenya, P. (2013). Situated knowledge and the EU sugar reform: a Caribbean life history, *Area*, 45(2), pp. 188–197.

Riggs, E.M. (2005). Field-based education and indigenous knowledge: essential components of geoscience education for Native American communities, *Science Education*, 89, pp. 296–313.

Rizvi, A.R., and van Riel, K. (2014). Nature based solutions for climate change adaptation – knowledge gaps, IUCN EbA Knowledge Series – Working Paper, Gland, Switzerland: IUCN. Retrieved from www.iucn.org/sites/dev/files/eba_knowledge_gaps.pdf [Last accessed on 10 November 2019].

Rose, N. (1999). *Powers of freedom: reframing political thought*, Cambridge: Cambridge University Press.

Rosendahl, J., Zanella, M.A., Rist, S., and Weigelt, J. (2015). Scientists' situated knowledge: strong objectivity in transdisciplinarity, *Futures*, 65, pp. 17–27.

Santha, S.D. (2007). State interventions and natural resource management: a study on social interfaces in a riverine fishery setting in Kerala, India, *Natural Resources Forum*, 31, pp. 61–70.

Santha, S.D. (2008). Local culture, technological change and riverine fisheries management in Kerala, South India, *International Journal of Rural Management*, 4(1&2), pp. 25–45.

Santha, S.D., Balasubramaniam, S., and Soletti, A.B. (2017). Interfaces in social innovation: an action research story on a tribal women's collective, *Glocalism: Journal of Culture, Politics and Innovation*, 3, DOI: 10.12893/gjcpi.2017.3.9 [Last accessed on 3 November 2019].

Schatzki, T.R. (2005). Peripheral vision: the sites of organizations, *Organization Studies*, 26(3), pp. 465–484.

Schau, H.J., Muñiz, A.M., Jr., and Arnould, E.J. (2009). How brand community practices create value, *Journal of Marketing*, 73, pp. 30–51.

Schetter, C. (2005). Ethnoscapes, national territorialisation, and the Afghan war, *Geopolitics*, 10(1), pp. 50–75.

Scott, J.C. (1998). *Seeing like a state: how certain schemes to improve the human condition have failed*, New Haven, CT: Yale University Press.

Seixas, C.S., and Berkes, F. (2003). Dynamics of social–ecological changes in a lagoon fishery in southern Brazil, in F. Berkes, J. Colding, and C. Folke (Eds.). *Navigating social-ecological systems: building resilience for complexity and change*, Cambridge: Cambridge University Press, pp. 271–298.

Sen, A. (1999). *Development as freedom*, New York: Oxford University Press.

Seur, H. (1992). The engagement of researchers and local actors in the construction of case studies and research themes, in N.E. Long and A. Long (Eds.). *Battlefields of knowledge: the interlocking of theory and practice in social research and development*, London: Routledge, pp. 115–143.

Sheridan, M.J., and Nyamweru, C. (Eds.). (2008). *African sacred groves: ecological dynamics and social change*, Athens, OH: Ohio University Press.

Sillitoe, P. (1998). The development of indigenous knowledge: a new applied anthropology, *Current Anthropology*, 39, pp. 223–252.

Sillitoe, P. (2010). Trust in development: some implications of knowing in indigenous knowledge, *Journal of the Royal Anthropological Institute*, 16, pp. 12–30.

Skoglund, A., and Jensen, T. (2013). The professionalization of ethics in the Intergovernmental Panel on Climate Change (IPCC) – from servant of science to ethical master? *Sustainable Development*, 21, pp. 122–130.

Smit, B., and Wandel, J. (2006). Adaptation, adaptive capacity and vulnerability, *Global Environmental Change*, 16(3), pp. 282–292.

Somekh, B. (1995). The contribution of action research to development in social endeavours: a position paper on action research methodology, *British Educational Research Journal*, 21(3), pp. 339–355.

Srang-iam, W. (2013). De-contextualized knowledge situated politics: the new scientific–local politics of rice genetic resources in Thailand, *Development and Change*, 44(1), pp. 1–27.

Stark, J.L. (2014). The potential of Dewyean-inspired action research, *Education and Culture*, 30(2), pp. 87–101.

Stoetzler, M., and Yuval-Davis, N. (2002). Standpoint theory, situated knowledge and the situated imagination, *Feminist Theory*, 3(3), pp. 315–333.

Sunil, B. (2019). Gender, body and the self: a study on the induced abortion experiences of women in Kumbakonam, Tamil Nadu, *Doctoral dissertation*, Tata Institute of Social Sciences, Mumbai, India.

Tandon, R. (1981). Participatory research in the empowerment of the people, *Convergence*, 14(3), pp. 20–27.

Tanner, T., and Seballos, F. (2012). Action research with children: lessons from tackling disasters and climate change, *IDS Bulletin*, 43(3), pp. 59–70.

Tengö, M., and von Heland, J. (2012). Adaptive capacity of local indigenous institutions: the case of the taboo forests of southern Madagascar, in E. Boyd and C. Folke (Eds.). *Adapting institutions: governance, complexity and social–ecological resilience*, Cambridge: Cambridge University Press, pp. 37–74.

Tickner, J.A. (1997). You just don't understand: troubled engagements between feminists and IR theorists, *International Studies Quarterly*, 41, pp. 611–632.

Verschoor, G. (1992). Identity, networks and space: new dimensions in the study of small-scale enterprise and commoditization, in N.E. Long and A. Long (Eds.). *Battlefields of knowledge: the interlocking of theory and practice in social research and development*, London: Routledge, pp. 171–188.

Wagenaar, H., and Cook, N. (2003). Understanding policy practises, in M. Hajer and H. Wagenaar (Eds.). *Deliberative policy analyses: understanding governance in the network society*, Cambridge: Cambridge University Press, pp. 139–171.

3   The landscape of adaptive innovation

Introduction

Climate change adaptation is one of the most recent, yet complex phenomenon that we, as social workers deal with in our practice arenas. In this book, climate change adaptation is referred to as those processes, actions and outcomes that could enable social–ecological systems or its elements to better cope with, manage or adjust to actual or expected climate and its effects. There is a fine line between what constitutes climate change adaptation and what actors generally construe as a 'conventional' development issue (de Wit, 2018, p. 38). Adhering to the dominant notion of treating climate change as a conventional development issue could result in the framing of adaptation as solely a technical problem (de Wit, 2018; Taylor, 2015). Therefore, we need to consistently be aware of how global ideas of adaptation encounter with the local conceptualisation of changes within the social–ecological system. To develop meaningful and resilient adaptation strategies, we have to recognise and work with knowledge systems and situated practices of local community actors who are impacted by climate change. This necessitates that our practice be rooted in the lived experiences of diverse vulnerable groups and their everyday struggles of interacting with a complex social–ecological system (Dennis et al., 2016; Ojha et al., 2014).

The aim of this book, as mentioned in the first chapter, is to introduce and elaborate an adaptive innovation model that could enable social workers and development practitioners to facilitate people-centred, fair and just adaptation mechanisms to climate change. Adaptive innovation involves a combination of observing, thinking, doing and reflecting wherein knowledge is acquired, shared and assimilated through iterative practices of different actors. It locates climate change adaptation processes as innovations that could shape both knowledge and practice of community actors in responding and adjusting to dynamic social–ecological systems. In most cases, these responses are linked to the everyday practices and livelihood struggles of community actors. Adaptive innovation strives towards the designing of people-centred adaptation strategies that will have the intent and effect of our decisions and actions on the structures and processes shaping equality and justice, care and empowerment for those who are disadvantaged and vulnerable in society. Such an approach therefore involves not only strengthening the adaptive capacities of vulnerable groups, but also deconstructing certain dominant narratives of adaptation (Eriksen et al., 2015; Ferdinand, 2019; Reed et al., 2015). As social workers, we can facilitate these processes by relying on the strengths of people, their communities, cultural narratives, pluralist knowledge systems and wider social networks (Peeters, 2012).

Adaptive innovations are situated reflective practices that are context specific, developmental and committed to the values of care, justice and solidarity. It emphasises on people-centred innovation processes by which local community actors collectively analyse their own situations in the context of social and ecological transitions; forge constructive partnership with other relevant actors to dialogue, ideate and develop working models; implement and critically observe, reflect and validate their adaptive strategies to the emergent contexts. The ultimate aim of adaptive innovation is to nurture caring solidarity, strengthen adaptive capacities, transform institutions as people-centred and enhance the resilience of social–ecological systems through collective decision-making and action. In the following section, the adaptive innovation model is presented by describing (a) the guiding values of justice, care and solidarity that are embedded within (b) the six phases of the adaptive innovation cycle and enriched by (c) reflective practice and (d) analysis of actor interfaces in adaptation. The adaptive innovation model can be conceptually represented as shown in Figure 3.1.

Values guiding adaptive innovation

The key foundation of the adaptive innovation model are the values guiding ethical adaptation, namely justice, care and solidarity. Ethical adaptation to climate change is possible only if we embed the values of justice, care and solidarity as key elements of our practice. While these values are discussed in detail in Chapter 4, these concepts are introduced below to enable the reader to comprehend the model of adaptive innovation as a whole.

Justice

Climate change undermines the capabilities of vulnerable groups to live with dignity (Hetherington and Boddy, 2013; Holland, 2012). Climate change risks also portray uncertainties over access to and control over one's own physical and social environment (Santha et al., 2016; Tester, 2013). Constrained by factors such as food and livelihood insecurity, people could become helpless and end up seeking outside support to deal with the impacts of climate change (Dominelli, 2012). In this regard, those people whose lives and livelihoods are most vulnerable to the consequences of climate change and who have contributed the least to its causes should receive preferential support (Mearns and Norton, 2010; Peeters, 2012). Nevertheless, present-day planned adaptation strategies tend to ignore the cultural, social, economic and political contexts shaping people's marginalisation and vulnerability. They tend to discount the

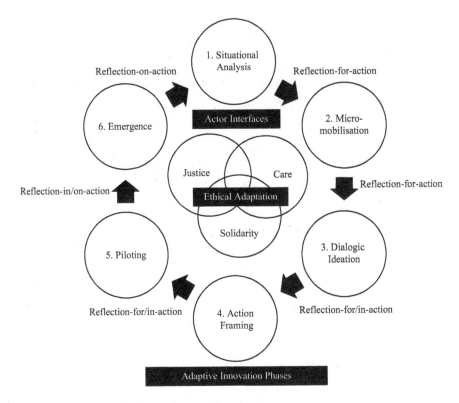

Figure 3.1 The adaptive innovation model.

underlying social inequalities and inequities that shape these diversities. Instead, on many occasions these inequalities are reinforced due to the insensitive top-down planned adaptation approaches that are laid out in the planned adaptation strategies (Klepp and Chavez-Rodriguez, 2018).

We have to certainly examine the nature of structural inequalities and the forces behind these inequalities to effectively implement the norms of ensuring equality (Carmalt, 2011; Fraser, 1995). Factors such as historical and contemporary forms of social exclusion compounded by non-inclusive economic and political ideologies lead to multiple dimensions of vulnerability (Reitan and Gibson, 2012; Santha et al., 2016; Schlosberg, 2012; Wisner et al., 2004). The poor and marginalised sections of the population often have limited access to institutions and resources, thus constraining their participation and decision-making capacities to mitigate risks or cope with it (Cleaver, 2005; Kwan and Walsh, 2015; Santha and Sunil, 2009; Wisner et al., 2004). These vulnerabilities also get reinforced and become even more complex in the context of other dynamic pressures such as deforestation, unemployment and neoliberal forms of development (Gotham, 2012; Tierney, 2015).

Any practice aiming at climate justice should strive towards the provision of rights, responsibilities and recognition (Bulkeley et al., 2014). Rights can be understood as the right to be protected from the impacts of climate change and from the costs of climate change actions, and to benefit from responses to climate change (ibid). It also brings to the forefront the right to participate and how this should be exercised (Bulkeley et al., 2013; Popke et al., 2014). Responsibility entails that the governing actors ensure these rights are guaranteed through both distributive and procedural mechanisms (ibid). Recognition demands and assures voice for the right to self-determination and dignified existence of different cultural and social groups in the face of climate change (Burnham et al., 2013). A single conceptualisation of a normative distributive justice cannot be therefore applied in the context of climate change adaptation (Young, 2003). In this regard, the historicity of marginalisation and resource inequities among those vulnerable to climate change can never be ignored (Whyte, 2019). Climate change adaptation has to be thus founded on inclusive, representative and consultative principles that recognise the situatedness and plurality of knowledge systems, which are capable of enabling vulnerable and marginalised groups to shape decision-making and become owners of their own development pathways (Morchain, 2018). Adaptive innovation is envisaged as a socially inclusive and people-centred process, where all participant actors are accepted and viewed as sufficiently capable to engage with the planning and designing of suitable and effective adaptation strategies. Among them, the most vulnerable groups should be recognised and represented to identify their valued beings and doings, interests and positions (Kronlid, 2014; Popke et al., 2014).

Ethics of care

Yet another important perspective that could facilitate ethical adaptation is the ethics of care, which emphasises on the moral significance embedded in people's routinised and situated relationships (Banks, 2008; Parrott, 2010). The ethics of care refer to 'approaches to moral life and community that are grounded in virtues, practices and knowledge associated with appropriate caring and care taking of self and others' (Whyte and Cuomo, 2017, p. 235). It emphasises on the relational dimensions of actors in specific social–ecological systems. It also highlights the affective foundation of ethical behaviour such as love, empathy and compassion, which could shape moral behaviour and inform actor's judgements about what is right and fair (Hoggett et al., 2009). Caring has to be viewed as a

> species activity that includes everything that we do to maintain, continue, and repair our 'world' so that we can live in it as well as possible. That world includes our bodies, ourselves, and our environment, all of which we seek to interweave in a complex, life-sustaining web.
>
> (Fisher and Tronto, 1991, p. 40; Tronto, 1993, p. 103)

Such a notion of caring highlights the web of reciprocal obligations and commitments that characterise relationships between diverse actors giving care and receiving care within a social–ecological system (Hoggett et al., 2009). The emphasis is both on the interdependence between human actors as well as between human and non-human actors.

Care is both a rational and affective engagement with the world (Tronto, 1993). The ethics of care values emotions such as sympathy, empathy, sensitivity and responsiveness (Held, 2006). According to Tronto (1993, p. 127), the four values that constitute an ethic of care are 'caring about, taking care of, care-giving, and care-receiving'. These four values of care are connected to our everyday life (Till, 2012). The practice of caring could be understood as an obligation to these values (Parrott, 2010; Tronto, 1993). As practice, care emphasises on 'our response to needs and why we should do so' (Held, 2006, p. 42). In this regard, four ethical elements of care are attentiveness, responsibility, competence and responsiveness (Tronto, 1993). All these elements are crucial towards addressing problems such as climate change, as it sees 'the interests of carers and cared-for as importantly intertwined rather than as simply competing, and fosters social bonds and cooperation' (Held, 2006, p. 15).

The ethic of care also showcases the relational and contextual nature of problems such as climate change adaptation and justice (Whyte and Cuomo, 2017). Any attempts towards fair and just adaptation should recognise the historical and intersectional drivers of exploitation and oppression of vulnerable groups, and the ethics of care and healing that these groups practice in their day-to-day life (Ranganathan and Bratman, 2019). As a transformative approach, 'it demands not only equality for subjugated actors in existing structures of society, but equal consideration for the experience that reveals the values, importance, and moral significance, of caring' (Held, 2006, p. 12). More importantly, the notion of caring recognises the subjectivity of human relationships, taking into account the needs of both self and others (Gilligan, 1982; Held, 2006; Parrott, 2010). Such a perspective also provides sufficient scope to recognise the situatedness and differences in the needs of others.

Ethical decision-making is situated in the quality of relationships that vulnerable people and dependent groups maintain with one another (Parrott, 2010). An ethic of care perspective can help in locating how these actors are situated and entangled in relations of injustice and carelessness (Williams, 2017). Building trust and empathy with these actors become crucial for ethical adaptation. Facilitating shared conversations that connect people and encourage them to openly share their ideas and are willing to listen to others' voices could help in building trust and sustaining caring relationships (Morchain, 2018). This necessitates that we have to take into account the lived experiences and emotions of all those within this network of relationships. Such relationships also need to be understood from a specific orientation that involves compassion or care within them (Parrott, 2010).

Solidarity

Adhering to fairness of procedures with a sense of caring and reflexive solidarity would make climate justice meaningful. Recognition in justice requires the recognition of social solidarity that emerges through collectively shared goals (Honneth, 1995). And this coexists with the relational networks based out of care and rights (ibid). Drawing insights from the works of Seyla Benhabib (2004) and Nancy Fraser (2008), Barnett (2011) quotes that the principles of justice must be interpreted dialogically and should acknowledge the situated contexts of social integration and solidarity among community actors. In this regard, 'justice depends on solidarity, on the feeling of being connected to others, … gives us a chance for meaningful participation, and respects our individual personalities even while giving us the feeling that we are all in the same boat' (Alexander, 2006, p. 13).

In a broader sense, solidarity could be understood as a concern or a situation in which the well-being of one person or group is positively related to that of others' (de Beer and Koster, 2009; McCoy, 2018). It is a moral value and social attitude linked to mutual respect, empathy and care, support and commitment to a shared cause (Arts et al., 2001; Reyes, 2016; ter Mullen et al., 2001). This notion of solidarity facilitates the provision of formal and informal care to vulnerable groups (ter Mullen et al., 2001). For many poor and marginalised communities, the only power they have is their local knowledge and their solidarity networks (Doherty and Doyle, 2013). And to be more specific, it is the marginalised and vulnerable groups who ought to have the first and primary say on human solidarity (UNDP, 2007). It is the solidarity amidst these actors that would facilitate 'civil repair', reflecting upon the capacity of these groups in the civil sphere to advance claims to power and respect (Alexander, 2006, p. 208).

There is a need for nurturing and strengthening both intergenerational and intersectional solidarity in climate change adaptation. Intergenerational solidarity is crucial for personal and social security, and for human bonding (Cruz-Saco and Zelenev, 2010). It helps in honouring and maintaining the strengths and abilities of each generation through mutual support, care and exchange of services (ibid). It also helps to 'bond people together through the values, associations and interactions, consensus and exchange, agreements, feelings of affective orientation, and similarities' (Cruz-Saco, 2010, p. 10). It also facilitates the transfer of knowledge among persons of all age group and is an outcome of social processes that are shaped by factors such as age, gender, culture and ethnic background, socioeconomic status and values along with certain emerging worldviews (ibid). Solidarity also has scope to include relations with non-human nature and future generations as well (Doherty and Doyle, 2013).

Problems associated with social–ecological systems have to be addressed in a more collaborative manner, where community actors in partnership with experts and local authorities define the nature and scope of the problem. Instead of simply providing some space for dialogue, there should be an element of active citizenship embedded in these processes such that local community actors participate in making changes at the ground level and have ownership over the governance processes of adaptation (Pant, 2014). Each actor should have a genuine respect for one another's contributions and nurture long-term shared commitments to address critical social–ecological concerns (Brydon-Miller, 2008). The inherent capacity of community actors to collectively respond to changing circumstances has to be identified, nurtured and enhanced. This implies that we have a crucial role in enabling local community actors to identify and leverage existing strengths as a solidarity group rather than imposing them from outside.

The phases of adaptive innovation

The six practice phases of the adaptive innovation cycle are (i) *Situational Analysis*, (ii) *Micro-mobilisation*, (iii) *Dialogic Ideation*, (iv) *Action Framing*, (v) *Piloting* and (vi) *Emergence*. Each phase of innovation is visualised as a process of discovering and reflecting on new ways of forging partnerships, nurturing participation and co-creating actionable solutions. These phases are interlinked and constitute a dynamic process of discovering new ways of participation, co-creation and situated learning. It demonstrates an iterative and reflective pathway to design diverse adaptation strategies suiting local contexts and knowledge frameworks embedded within a complex social–ecological system. Each of these phases are briefed below.

1. **Situational analysis**

   All the key elements in the social–ecological system and their interrelations, and the discursive positions and practices of actors in which the complex phenomena of livelihood security, climate risks and uncertainties are embedded constitute the situation. The meanings attached to the experiences arising out of diverse social encounters also characterise the situation. Situational analysis is the process of understanding the vulnerability contexts, livelihood practices, adaptation trends and other key issues affecting diverse community actors in a given social–ecological system. In a deeper sense, it seeks to analyse 'a particular situation of interest through the specification, re-representation and subsequent examination of the most salient elements in the situation and their relations' (Clarke, 2005, p. 29). Some of the important steps that could guide the situational analysis are scoping review, participatory mapping and analysis of drivers and barriers to ethical adaptation.

2. **Micro-mobilisation**

   Micro-mobilisation is a strategic process aimed at organising community actors to participate collectively through innovation platforms in devising suitable adaptation strategies. It involves a range of interactive processes that are designed to mobilise or influence diverse intersectional community actors towards creating a shared vision and acting upon it (Snow et al., 1986). It also involves motivating actors and building consensus to harness collective support, structure and momentum for action (Klandermans, 1984; McEntire et al., 2015). It is envisaged that members of the innovation platform will participate through a genuine and emotional commitment to the shared vision, engage proactively in consensus building and design possible adaptation actions. These are in itself the ultimate outcomes of the micro-mobilisation processes (McEntire et al., 2015; Ward, 2016). The micro-mobilisation processes are supported by steps such as shared visioning, and risk and vulnerability analysis.

3. **Dialogic ideation**
   The third phase of adaptive innovation emphasises on the multiplicities of imagination and narratives of actors' lived reality. Dialogic ideation is a process collective imagination where community actors in partnership with other stakeholders attempt to ideate and co-create multiple adaptation pathways through in-depth deliberations, dialogue and other forms of shared conversations and decision-making. It also involves exploring several possibilities that could generate new thinking towards realising those options (Southern, 2015).

4. **Action framing**
   Through the previous phases, we would have facilitated the participant actors to gather an in-depth understanding of their situations as well as explore together various possibilities and ideas to address issues of their concern. Action framing refers to the collective and participatory processes that involve translating the emergent ideas into meaningful sensory experiences, images or visuals of action. It allows for systemic exploration of actionable possibilities. The aim of action framing is to explore the feasibility and resilience of broader ideas, generate shared experiences, identify resources and foresee the nuisances of implementing the idea into an actionable model.

5. **Piloting**
   Piloting refers to iterative and reflective processes aimed at implementing and testing the suitability, feasibility and effectiveness of working models. Ideas and working models require constant reflection and refinement through practice. Pilot projects can enable participant actors to test their ideas, iterate and ascertain the feasibility of their working models. Pilots help us to take our ideas and models to the field at a faster pace, rather than spending too much time and resources in developing detailed plans and strategies. Pilots also help us to foresee crucial challenges in strengthening adaptive capacities as well as in being alert to both intended and unintended consequences of our adaptation efforts. If agreed upon by the community actors, the piloted projects after further considerable iterations and reflections can be scaled up for wider implementation.

6. **Emergence**
   Emergence refers to the emerging self-organised patterns of systemic arrangements, which are both the intended and unintended consequences of the adaptation project. Emergence is an integral part of any dynamic social–

ecological system. There can be surprises in climate change adaptation. These surprises may manifest in terms of both intended and unintended consequences. Emergence is the creation of something new, which is qualitatively different from the phenomena out of which it has emerged (Capra, 2002). This implies that the characteristics of both the structures and outcomes of the emergence phase could be entirely different from the characteristics of the previous adaptive innovation phases. The phase of emergence can result in a new order in social, ecological and cognitive arenas (ibid). Subsequently, it could result in the co-creation of new knowledge, skills and ideas. New members, additional materials and ideas could be added into emergent structure and blend into the social–ecological system. If the system attains a breakthrough, we could say that our adaptation efforts have reached a new state of order in a specific temporal-spatial context. Or else the system would have faced a breakdown indicating that our adaptation efforts have either failed or turned out to be maladaptive. In either of these cases, adaptive innovation would follow a reflective cycle and move again towards next levels of situational analysis. All the above-mentioned phases of adaptive innovation are elaborated in detail in Chapter 5 of this book.

Reflective practice

Reflective practice can be understood as iterative cycles of action-reflection, where reflection on action taken at each phase of adaptive innovation would set the stage for the next phase. The adaptive innovation model draws inspiration from action research and reflective practice traditions. If climate change is an inconvenient truth (Gore, 2006), then adaptive innovation is a quest for truthful reflection and action. The quest should be looked upon as an iterative cum reflective process that involves dealing with uncertainty and surprises at different levels of the social–ecological system (Berkes et al., 2003; Holling, 1978; Walters, 1986). It, therefore, involves 'repeated cycles of examining practice, adjusting practice and reflecting upon it, before trying it again' (Grushka et al., 2005, p. 239). The whole process is also a kind of collective and collaborative learning where participant actors co-evolve adaptation strategies by thinking, doing and reflecting on the doing. Each phase of adaptive innovation is conceived as an impetus to create involvement and generate iterative feedbacks for further reflection and action.

The concept of 'reflective practice' gained its prominence through an iterative and experimental learning approach developed by Donald Schön (1983, 1987). Experimental learning in this context is not a process of trial and error. Instead, it is a cycle of action-reflection, where reflection on each effort taken sets the stage for the next attempt (Schön, 1987). It involves the shaping of strategic action and generation of theoretical understanding through a series of successive triangulations (ibid). Reflective practice includes the ability to work across differing world views without promoting oppressive practice, and at the same time gather insights into one's own situations of practice and discover possibilities to create alternative frames of reality (Adamowich et al., 2014; Schön, 1983; Zambo, 2014). It equips each participant actor to reinterpret his or her own values, interests and knowledge with respect to those of others (Elliott, 1991; Norton et al., 2011).

Critical reflexivity also involves questioning our own positionalities, assumptions and actions as social workers (Maxey, 2004). Reflexivity is thus a key element that would help us to understand our own performance and practice (Postholm and Skrovset, 2013). Our capacity for reflexivity in each situation of practice is thus central to our actions, decisions and judgements made (Brown et al., 2005). Reflection also helps us to develop context specific theories that could inform our future practice (Killion and Todnem, 1991). Reflection in this context has to be looked upon as 'a process that encompasses all time designations, past, present, and future simultaneously' (Killion and Todnem, 1991, p. 15). As part of the adaptive innovation model, this book presents a three-phased reflection cycle, namely *reflection-for-action*, *reflection-in-action* and *reflection-on-action*. Reflection-for-action is thinking about future actions with the intention of improving or changing a practice (Grushka et al., 2005; Killion and Todnem, 1991; Olteanu, 2017). It is a kind of future scenario making, which includes elements of thinking about practice in order to improve it (Olteanu, 2017; Ong, 2011). It emphasises on the need to rely on multiple perspectives of the self, other participant actors, co-social workers and the available documents including literature (Ong, 2011). In the adaptive innovation model, reflection-for-action is crucial during the phases of situational analysis, micro-mobilisation and dialogic ideation. In this regard, the situation per se becomes the ultimate unit of our critical reflection. We need to explore and critically reflect on the various elements associated with the situation including diverse community actors and their social positions and its intersectional dimensions with livelihood risks and uncertainties, vulnerabilities, capacities and power relations. The experiences, needs and aspirations can be historically and structurally different for diverse community actors and through reflection-for-action, we along with other participant actors will be in a process of refining, defining and getting clarity on strategizing future action.

Reflection-in-action takes place during an action, and reflection-on-action takes place after an event has occurred (Schön, 1983, 1987). Reflection-in-action is reflection on actors' situatedness and being alert to simultaneous changes in the social–ecological system. It is an iterative way of thinking and acting in the midst of action (Killion and Todnem, 1991; Schön, 1983). To commence reflection-in-action, one needs to explore the nature of routinised responses that community actors bring to their day-to-day practice situations (Schön, 1985). It consists of 'strategies of action, understanding of phenomena, and ways of framing the problematic situations encountered in the day-to-day experiences' of actors (Schön, 1985, p. 24). Mostly, reflection-in-action is a tacit and dynamic process. There is a need for nurturing the reflection-in-action element, where reflection tends 'to focus not only on the outcomes of action, instead on the action itself and the intuitive knowing implicit in the action' (Schön, 1983, p. 56). Reflection-in-action is like contemplating each social encounter and thereby our own practice (Norton et al., 2011). With respect to the adaptive innovation model, the reflection-in-action processes predominantly features during dialogic ideation, action framing and piloting phases. Reflection-on-action, on the other hand, is reflection on practice and on one's actions and thoughts that follow the implementation of the adaptation strategy. It is a critical account of what really happened and is still continuing to happen or emerging due to the action. It is also an exploration of what were the intended and unintended consequences of one's action; and which aspect of this outcome is that one is able to explain, and one is not able to comprehend. It largely features during the piloting and emergence phase of the adaptive innovation process.

In a continuous spiral of action-reflection, reflection-for-action is the desired outcome of both reflection-on-action and reflection-in-action (Killion and Todnem, 1991). It is through a deeper introspection of one's past and present actions, community actors will be able to generate knowledge that will inform their future courses of action (ibid). Not giving due emphasis to all three phases of reflective practice could discount or over-simplify need analysis and planning for adaptation (Thompson and Pascal, 2012). This also implies that we should avoid going to the communities with pre-determined notions of what the adaptation strategy or innovation design will be. Instead we should try to embrace and support the needs, constraints and local knowledge of our participant actors (Maxey, 2004). It is advisable that we design and maintain few learning and actionable questions at the beginning and culmination of each phase of the adaptive innovation process. These lines of inquiry will help us to systematic- ally nurture, analyse and document the whole process through the lens of reflective practice. Field diaries and journals could also act as important medium to systematically pursue the path of reflective practice.

Each social encounter characterises a dynamic interplay of observing and listening, thinking and doing. Reflection in and on action include how in each social encounter, different community actors behave and respond to the values, interests and knowledge of other actors. Focusing on specific social encounters could help us in challenging basic understandings and be empath- etic to the experiences, vulnerabilities and capacities of those subjugated com- munity actors, who are often marginalised to the needs of the dominant interest groups. This could also create adequate spaces and capacities for people at the margins to develop new, empowering meanings of their actions (Thompson and Pascal, 2012). The emphasis is on the creative struggles of community actors located in specific social positions as they engage with the practicalities of their own social and historical conditions (Clarke, 2005). This further implies that we have to be more sensitive and empathetic, as we enter the life-worlds of community actors. A deeper and enhanced understanding of actor's behaviour, actions and interactions is crucial to analyse social change in the context of climate change adaptation. In this regard, the actor interface approach discussed in the subsequent section could provide us an analytical pathway to explore the dynamics of specific social encounters in a deeper and reflective manner.

Analysing actor interfaces in adaptation

Community-based adaptation strategies need to be contextualised and reflected upon as interfaces between diverse actors in different relations of power, whose outcome depends upon the capacities, tactics and rationality of respective actors in making adaptation decisions (Santha, 2015). Any social encounter in such practice contexts have to be looked upon as situations in which actors with different rationalities and power relations interface with each other (Long, 2001). Actors in this context are not viewed as passive recipients of adaptation interventions. Instead, they are recognised as active stakeholders with specific knowledge and capacity for action. Actors are thus both 'knowledgeable' and 'capable', who 'enter into each specific social situation with new capacities to act and negotiate' (Gerharz, 2018, p. 2). In this book, actor interfaces could be understood as critical points of interaction located in specific social encounters that involve multiple actors with diverse values, interests, knowledge and power; and the consequences could be either social innovations or social discontinuities.

Each phase of adaptive innovation will be characterised by multiple and often complex interpretations of reality and meanings attributed by diverse community actors who are involved in the situation. Every social encounter could be shaped by the micro-politics of power showcasing the submissions and resistances of diverse actors (Clarke, 2005). The recognition that there is competing knowledge and interest in any given situation also helps us to remain alert to power relations and ideological issues interfacing with practice (Kirkwood et al., 2016). If not nurtured sensitively, these social encounters can become sites of domination and spaces for the perpetuation of mere socio-technical interest. We need to therefore consistently reflect on aspects such as language and narratives, negotiations and power relations, embedded in these social encounters (Thompson and Pascal, 2012). The uniqueness of each social encounter, the complexities involved in decision-making and the professional anxieties of us practitioners also become important factors shaping adaptation (Ruch, 2002).

The analysis of interrelationship and interdependency between structures and action is a key focus of the social interface approach. Social interfaces could be understood as a 'critical point of intersection between different life worlds, social fields or levels of social organisation, where social discontinuities based upon discrepancies in values, interests, knowledge and power are most likely to be located' (Long, 2001, p. 177). The critical points of interaction are those social encounters where community actors confront each other; and struggles over social meanings and practices takes place (Long and Long, 1992). These interactions are often weaved around the problems of designing alternate ways of negotiating, bridging, accommodating, withdrawing or struggling against each other's different social and cognitive worlds (Long, 1989, 2001). Interface analysis thus helps us to understand the dynamic, emergent and conflictive nature of social encounters between diverse actors with differing interests, knowledge, resources and power (Long, 1989; Long and Long, 1992).

The interface approach gives emphasis to the specifics of social contexts characterising every social encounter. Each social encounter would involve the interaction between many different actors with their own specific knowledge, values, interests and concerns. And the actions of a particular actor may not therefore be always compatible with the interests of another (Gerharz, 2018; Long, 2001). The efficacy of the interface approach is that it enables us to focus on the negotiations and mediations that takes place at each phase of the adaptive innovation project. As we comprehend these behaviours and strategies of actors, we will also be able to unearth the rationalities shaping these actions and how they are articulated in specific ways. Understanding the dynamics of these social encounters will also help us to understand why some of the adaptation strategies that were designed and piloted have proven to be unsuccessful or resulted in unintended consequences. The social interface analysis could also provide insights on those social encounters that were strategic in terms of how diverse actors agreed upon, negotiated or contested while arriving at decisions pertaining to adaptation. It also helps us to interrogate the diverse situations that led to the emergence of actionable knowledge and new validations for action. We could also reflect on how shared understandings and conflicts emerged with varying contexts across space and time. During each phase of the adaptive innovation process, we could facilitate community actors to collectively reflect on these contexts and their dynamics and sustain meaningful thinking and doing to deal with these situations (Colucci-Gray and Camino, 2016). Further insights on interface analysis is provided in Chapter 7 of this book.

S.F.F.3.1 The story of a cattle pond.

In the late 1990s, I worked with a non-governmental organisation that strives towards the ecological restoration and conservation of common property resources in India. I was posted in Madanapalle that had its own unique social–ecological features. Surrounded by the Deccan Plateau, the Eastern Ghats and the Western Ghats, this region is marked by scattered hill ranges, vast stretches of common lands (wastelands) and forests, numerous chains of rainfed irrigation tanks and a very distinctive biodiversity. The communities were largely involved in rain-fed agriculture. I was working with communities that were highly vulnerable to drought, severe water scarcity, out-migration and distress sale of both crops produce and livestock. In these communities, marginal and small farmers who did not have access to irrigation were largely dependent on groundnut and millet cultivation. And those farmers who had access to irrigation cultivated tomato, sunflower, sugarcane, vegetables, jasmine and groundnuts. These villages were marked by severe caste and gender-based inequalities as well. Landholding farmers were largely from privileged caste groups, while the landless farmers, labourers and pastoralists belonged to highly deprived and socially excluded caste groups. The women, especially single women in these communities were more vulnerable due to lack of livelihood resources and also due to other forms of social and economic exclusion that were shaped by intersectional patriarchal structures. Many women in these communities did not have direct access to banks, hospitals, schools and the market.

I was part of a team that strived towards strengthening village level institutions in these communities to restore, manage and govern common property resources such as wastelands, degraded forest lands, grazing lands and irrigation tanks. Our efforts were largely guided by a ridge-to-valley-based watershed development approach, which also presumed that these attempts will ensure the livelihood security of many poor and marginalised households in these communities. Before selecting a village community for our intervention, we used to carry out a preliminary situational analysis in consultation with community actors. Followed by which, we used to carry out transect walks, participatory mapping and stakeholder workshops to draw relevant insights on the everyday experiences and aspirations of community actors.

Gradually, the community members were mobilised to form a Tree Growers' Mutually Aided Cooperative Society (TGMACS). We used to rely on traditional oral storytelling techniques such as the '*Hari Katha*' or the '*Burra Katha*' to mobilise community actors. The storytelling team used to consist of one main performer and two co-performers. These orators were largely local community members from our other project villages, who were skilled in communicating the project vision and goals by twinning it with their own belief systems and lived experiences. This three to four-hour narrative entertainment format also weaved in other folk elements such as solo drama, dance, songs, poems, prayers and jokes. The topic was largely based on a mythological story that equally reflected certain contemporary social-ecological realities of the community. These storytelling practices also contextualised and situated forests, climate and natural resources within the everyday lived experiences of the audience and emphasised on communities' role in ecological restoration and conservation.

The members of the TGMACS used to develop their village level perspective plans in consultation with our project team. Usually, the male members of the cooperative used to attend these planning meetings. Gradually, through our facilitation, women also began to attend such meetings. The ideas and action plans that were developed was largely based on certain pre-conceived technocratic notions of watershed development, which included planning for activities such as check dams, gully plugs, contour trenching and afforestation works. Though the consultations seemed to be 'participatory' towards developing the micro-plans, the discussions often ended up in choosing specific activities from a pre-existing list of various soil and water conservation techniques. People were eager to participate in such projects, as it ensured regular wage employment during times of drought and distress migration.

One night, my planning meeting with a group of villagers got extended beyond the normal duration. As there was no public transport service available, I decided to stay back in the village for that night. Next day early morning, I walked towards the bus stop that was located across a dried-up riverbed. As I was crossing the riverbed, I saw an elderly woman (almost in her late 70s) grazing her goats and sheep along the riverbed. I had never seen her before in the village meetings. I observed that she was walking towards me and therefore I stopped and began interacting with her. Her name was Mallamma, a common name among women in the region. It is a synonym to the name of the goddess who is believed to reside in their forests. Mallamma was a widow pastoralist and belonged to a less-privileged caste. She used to leave her village at dawn with her cattle, sheep and goats and return by dusk. Being a single woman, she was seldom invited by other community members to attend any public meetings.

Mallamma asked me: 'You are the "mountain worker" right?'. (Mountain worker was the term given by local communities to practitioners like us who worked with communities dependent on forests and hill ranges in the region.)

I nodded my head, as if it was a 'yes'.

After a brief silence, Mallamma continued,

> I know that you are all doing good work to restore our forests and hills. It also gives our people employment within the village. But I have a request … Can you also help us in digging out a cattle pond in our village?! That will be really helpful for all our cattle, sheep and goats. Otherwise we have to walk many miles away to fetch water!

I was quite surprised and baffled with her request. I had never heard about such a concept of 'cattle pond' before! None of the watershed training programmes that I had attended did mention about such a technique. None among the villagers present in our earlier meetings had ever raised such a request. Unable to get out of that state of perplexity, I asked her,

'Our planning document does not have any provision for a cattle pond. But tell me, in this water scarce region, where will you dig a cattle pond? And what is the surety that we will find water?!'

In response, Mallamma pointed her quivering hands towards a greener pasture located adjacent to the riverbank. Though the riverbed was completely dry, that stretch of land was distinct from the rest due to the overgrowth of a specific variety of grass, locally called as 'jammu gaddi' (a kind of Elephant Grass). She then said: 'Wherever these plants are there, there will be water beneath it. If we dig below this, we will get water. But it requires labour, money and collective effort!' She stopped and looked eagerly into my eyes for a favourable response.

I replied, 'I will explore about this. Nevertheless, don't have any hopes, as I am not sure of the outcome'.

I was still perplexed at her understanding of the local ecology, as the dominant knowledge systems that were prevailing then did not recognise these needs and ideas. Mallamma returned to her work, and I continued walking towards the bus stop. It was then I felt an inner urge prompting me to return immediately to the village and discuss with other community actors about the conversation that I had with Mallamma. Without boarding the bus, I walked back to the village. People were surprised to see me back so soon. Upon their enquiry, I told them about the need and ideas that Mallamma shared with me. To my surprise, the whole village immediately called for an urgent meeting. One member from each of the 150 households participated in the meeting. Even Mallamma was asked to attend the meeting.

In the meeting, they unanimously said,

> Water for our cattle, sheep and goats is the foremost need. If you could help us in building a cattle pond, that would be the greatest support you can do for us and our cattle. If you could provide us with the material cost, we are even ready to contribute our labour as '*shramdaan*' (voluntary labour).

I asked them, 'Why did you not mention about this need ever before?'

An elderly member of the TGMACS replied, 'We did not ... because your technical guidelines for watershed development did not have any provisions for cattle ponds! We chose from the options that you gave us and not from our needs and knowledge!'

Later in the day, I discussed about these developments to my Team Leader and other staff members in the project. None among my team members agreed to the idea of a cattle pond. They believed that we should not take up any activity, if it is not prescribed in the watershed management guideline. After considerable negotiation, my Team Leader agreed to my request saying,

> Though we do not have provisions for a cattle pond, we have to recognise the significance of local communities' needs and their knowledge. You can go ahead with the plan of constructing the cattle pond as a pilot project. However, take care that you incur a very minimum cost.

I returned to the village, the very next day. The villagers were so happy to ideate and design a blueprint of the proposed cattle pond. A miniature working model of the cattle pond was built using clay and bamboo. After taking feedbacks from experts within the community, they modified the design. The cost of the whole pilot project was then estimated. As agreed, the villagers decided to contribute a part of their labour as '*shramdaan*'. This also enabled us to limit the project cost. They began the construction of the cattle pond at the same site that Mallamma referred to me earlier. Both men and women participated enthusiastically in the labour work. By the end of the sixth day, they found water. It took two more days to complete the construction. Then on the ninth day, they inaugurated the cattle pond by letting Mallamma's sheep and goats drink water from the pond. It was a moment of excitement and recognition for these people. Throughout that summer, this pond served water for the livestock of the neighbouring villages as well.

The most surprising aspect was that in the course of the next one year, around 33 villages, in my project area, built cattle ponds using their own knowledge and resources. Some among these villages did not even seek outside support to construct these ponds. And for my organisation, digging cattle ponds became an entry point activity. I left the organisation after two more years. Twenty years later, I came to know that the people in the village of Mallamma still narrate the story of their cattle pond to any outsider who goes to their village. Studies also mention that the revival of such small cattle ponds have had significant impact on the retention of moisture and increased availability of water and fodder for cattle in the project villages (Sinha, 2006).

Mallamma is no more. But she taught me important lessons on people-centred and inclusive practice. This book is a reflective outcome of these experiences. Nevertheless, the contexts of practice have undergone considerable changes. Today, it is not just a story of drought and water scarcity. The environment has become more complex with climate variability, knowledge uncertainties and livelihood insecurities. These transitions and emerging complexities reiterate the significance of recognising past experiences and local knowledge systems of community actors.

The aim of this chapter was to introduce the landscape of the adaptive innovation model. Adaptive innovation happens in a systemic space where ideas, knowledge, techniques and practice, guiding action and reflection of all concerned actors are distinct, but interwoven. It involves diverse community actors responding to the impacts of climate change through their own unique, but shared forms of knowing and practice. Local community actors would have their own understanding of climate change impact and response strategies. Similarly, external experts including us social workers would construct our own specific views of action, interaction and practice. The values, interests, knowledge positions and the spatial flow of power between the local community actors and outsiders shape the diverse social encounters and adaptation behaviour. This chapter was positioned from this understanding of change and practice. The next chapter elaborates how ethical adaptation to climate change based on the values of justice, care and solidarity is a crucial foundation for adaptive innovation.

References

Adamowich, T., Kumsa, M.K., Rego, C., Stodddart, J., and Vito, R. (2014). Playing hide-and-seek: searching for the use of self in reflective social work practice, *Reflective Practice*, 15(2), pp. 131–143.

Alexander, J.C. (2006). *The civil sphere*, New York: Oxford University Press.

Arts, W., Muffels, R., and ter Mullen, R. (2001). Epilogue, in R. ter Mullen, W. Arts, and R. Muffels (Eds.). *Solidarity in health and social care in Europe*, London: Springer-Science + Business Media, B.V., pp. 463–477.

Banks, S. (2008). Critical commentary: social work ethics, *British Journal of Social Work*, 38(6), pp. 1238–1249.

Barnett, C. (2011). Geography and ethics: justice unbound, *Progress in Human Geography*, 35(2), pp. 246–255.

Benhabib, S. (2004). *The rights of others: aliens, residents, and citizens*, Cambridge: Cambridge University Press.

Berkes, F., Colding, J., and Folke, C. (2003). *Navigating social-ecological systems: building resilience for complexity and change*, Cambridge: Cambridge University Press.

Brown, K., Fenge, L., and Young, N. (2005). Researching reflective practice: an example from post-qualifying social work education, *Research in Post-compulsory Education*, 10(3), pp. 389–402.

Brydon-Miller, M. (2008). Ethics and action research: deepening our commitment to principles of social justice and redefining systems of democratic practice, in P. Reason and H. Bradbury (Eds.). *The Sage handbook of action research: participative inquiry and practice*, 2nd edition, New Delhi: Sage, pp. 199–210.

Bulkeley, H., Carmin, J., Broto, V., Edwards, G., and Fuller, S. (2013). Climate justice and global cities: mapping the emerging discourses, *Global Environmental Change*, 23, pp. 914–925.

Bulkeley, H., Edwards, G., and Fuller, S. (2014). Contesting climate justice in the city: examining politics and practice in urban climate change experiments, *Global Environmental Change*, 25, pp. 31–40.

Burnham, M., Radel, C., Ma, Z., and Laudati, A. (2013). Extending a geographic lens towards climate justice, part 1: climate change characterization and impacts, *Geography Compass*, 7(3), pp. 239–248.

Capra, F. (2002). *The hidden connections: integrating the biological, cognitive, and social dimensions of life into a science of sustainability*, New York: Doubleday.

Carmalt, J.C. (2011). Human rights, care ethics and situated universal norms, *Antipode*, 43 (2), pp. 296–325.

Clarke, A.E. (2005). *Situational analysis: grounded theory after the postmodern turn*, New Delhi: Sage.

Cleaver, F. (2005). The inequality of social capital and the reproduction of chronic poverty, *World Development*, 33(6), pp. 893–906.

Colucci-Gray, L., and Camino, E. (2016). Looking back and moving sideways: following the Gandhian approach as the underlying thread for a sustainable science and education, *Visions for Sustainability*, 6, pp. 23–44.

Cruz-Saco, M.A. (2010). Intergenerational solidarity, in M.A. Cruz-Saco and S. Zelenev (Eds.). *Intergenerational solidarity: strengthening economic and social ties*, New York: Palgrave Macmillan, pp. 9–34.

Cruz-Saco, M.A., and Zelenev, S. (2010). Introduction, in M.A. Cruz-Saco and S. Zelenev (Eds.). *Intergenerational solidarity: strengthening economic and social ties*, New York: Palgrave Macmillan, pp. 1–6.

de Beer, P., and Koster, F. (2009). *Sticking together or falling apart? Solidarity in an era of individualisation and globalisation*, Amsterdam: Amsterdam University Press.

de Wit, S. (2018). A clash of adaptations: how adaptation to climate change is translated in northern Tanzania, in S. Klepp and L. Chavez-Rodriguez (Eds.). *A critical approach to climate change adaptation: discourses, policies and practices*, London: Routledge-Earthscan, pp. 37–54.

Dennis, M., Armitage, R.P., and James, P. (2016). Social-ecological innovation: adaptive responses to urban environmental conditions, *Urban Ecosystems*, 19, pp. 1063–1082.

Doherty, B., and Doyle, T. (2013). *Environmentalism, resistance and solidarity: the politics of Friends of the Earth International*, Hampshire: Palgrave Macmillan.

Dominelli, L. (2012). *Green social work: from environmental crises to environmental justice*, Cambridge: Polity Press.

Elliott, J. (1991). *Action research for educational change*, Bristol: Open University Press.

Eriksen, S.H., Nightingale, A.J., and Eakin, H. (2015). Reframing adaptation: the political nature of climate change adaptation, *Global Environmental Change*, 35, pp. 523–533.

Ferdinand, T. (2019). *Kenyan herders are switching from cattle to camels to adapt to climate change*, World Resources Institute, 9 July 2019. Retrieved from www.wri.org/blog/2019/07/kenyan-herders-are-switching-cattle-camels-adapt-climate-change [Last accessed on 17 July 2019].

Fisher, B., and Tronto, J.C. (1991). Toward a feminist theory of caring, in E. Abel and M. Nelson (Eds.). *Circles of care: work and identity in women's lives*, Albany, NY: State University of New York Press, pp. 36–54.

Fraser, N. (1995). From redistribution to recognition? Dilemmas of justice in a 'post-socialist' age, *New Left Review*, 212, pp. 68–93.

Fraser, N. (2008). *Scales of justice: reimagining political space in a globalising world*, Cambridge: Cambridge University Press.

Gerharz, E. (2018). *The interface approach*, IEE Working Papers, No. 212, Bochum: Ruhr-Universität Bochum, Institut für Entwicklungsforschung und Entwicklungspolitik (IEE). Retrieved from www.econstor.eu/bitstream/10419/183566/1/wp-212.pdf [Last accessed on 28 July 2019].

Gilligan, C. (1982). *In a different voice*, Cambridge, MA: Harvard University Press.

Gore, A. (2006). *An inconvenient truth: the planetary emergency of global warming and what we can do about it*, New York: Rodale.

Gotham, K.F. (2012). Disaster Inc: privatization and post-Katrina rebuilding in New Orleans, *Perspectives on Politics*, 10(3), pp. 633–646.

Grushka, K., McLeod, J.H., and Reynolds, R. (2005). Reflecting upon reflection: theory and practice in one Australian University teacher education program, *Reflective Practice*, 6(2), pp. 239–246.

Held, V. (2006). *The ethics of care: personal, political and global*, New York: Oxford University Press.

Hetherington, T., and Boddy, J. (2013). Ecosocial work with marginalised populations: time for action on climate change, in M. Gray, J. Coates, and T. Hetherington (Eds.). *Environmental social work*, London: Routledge, pp. 46–61.

Hoggett, P., Mayo, M., and Miller, C. (2009). *The dilemmas of development work: ethical challenges in regeneration*, Bristol: The Policy Press.

Holland, B. (2012). Environment as meta-capability: why dignified human life requires a stable climate system, in A. Thompson and J. Bendik-Keymer (Eds.). *Ethical adaptation to climate change: human virtues of the future*, Cambridge, MA: The MIT Press, pp. 145–164.

Holling, C.S. (1978). *Adaptive environmental assessment and management*, Chichester, UK: John Wiley and Sons.

Honneth, A. (1995). *The struggle for recognition: the moral grammar of social conflicts*, Cambridge, MA: The MIT Press.

Killion, J., and Todnem, G. (1991). A process for personal theory building, *Educational Leadership*, 48(6), pp. 14–16.

Kirkwood, S., Jennings, B., Laurier, E., Cree, V., and Whyte, B. (2016). Towards an inter-actional approach to reflective practice in social work, *European Journal of Social Work*, 19(3–4), pp. 484–499.

Klandermans, B. (1984). Mobilization and participation: socio-psychological expansions of resource mobilization theory, *American Sociological Review*, 49(5), pp. 583–600.

Klepp, S., and Chavez-Rodriguez, L. (2018). Governing climate change: the power of adaptation discourses, policies, and practices, in S. Klepp and L. Chavez-Rodriguez (Eds.). *A critical approach to climate change adaptation: discourses, policies and practices*, London: Routledge-Earthscan, pp. 3–34.

Kronlid, D.O. (2014). *Climate change adaptation and human capabilities: justice and ethics in research and policies*, New York: Palgrave Macmillan.

Kwan, C., and Walsh, C.A. (2015). Climate change adaptation in low-resource countries: insights gained from an eco-social work and feminist gerontological lens, *International Social Work*, 58(3), pp. 385–400.

Long, N. (1989). Conclusion: theoretical reflections on actor, structure and interface, in N. Long (Ed.). *Encounters at the interface: a perspective on social discontinuities in rural development*, Wageningen: Agricultural University Wageningen, pp. 221–243.

Long, N. (2001). *Development sociology: actor perspectives*, London: Routledge.

Long, N., and Long, A. (Eds.). (1992). *Battlefields of knowledge: the interlocking of theory and practice in social research and development*, London: Routledge.

Maxey, L.J. (2004). Moving beyond from within: reflexive activism and critical geographies, in D. Fuller and R. Kitchin (Eds.). *Radical theory/critical praxis: making a difference beyond the academy?* Vernon and Victoria, BC, Canada: Praxis (e)Press, pp. 159–171.

McCoy, B. (2018). The job guarantee: an institutional adjustment toward an inclusive provisioning process, in M.J. Murray and M. Forstater (Eds.). *Full employment and social justice: solidarity and sustainability*, Binzagr Institute for Sustainable Prosperity, Cham, Switzerland: Palgrave Macmillan, pp. 239–254.

McEntire, K.J., Leiby, M.J., and Krain, M. (2015). Human rights organisations as agents of change: an experimental examination of framing and micromobilisation, *American Political Science Review*, 109(3), pp. 407–426.

Mearns, R., and Norton, A. (2010). Equity and vulnerability in a warming world: introduction and overview, in R. Mearns and A. Norton (Eds.). *Social dimensions of climate change: equity and vulnerability in a warming world*, Washington, DC: World Bank, pp. 1–44.

Morchain, D. (2018). Rethinking the framing of climate change adaptation: knowledge, power, and politics, in S. Klepp and L. Chavez-Rodriguez (Eds.). *A critical approach to climate change adaptation: discourses, policies and practices*, London: Routledge-Earthscan, pp. 55–73.

Norton, C.L., Russell, A., Wisner, B., and Uriarte, J. (2011). Reflective teaching in social work education: findings from a participatory action research study, *Social Work Education*, 30(4), pp. 392–407.

Ojha, H.R., Sulaiman, R., Sultana, P., Dahal, K., Thapa, D., Mittal, N., Thompson, P., Bhatta, G.D., Ghimire, L., and Aggarwal, P. (2014). Is South Asian agriculture adapting to climate change? Evidence from the indo-gangetic plains, *Agroecology and Sustainable Food Systems*, 38, pp. 505–531.

Olteanu, C. (2017). Reflection-for-action and the choice or design of examples in the teaching of mathematics, *Mathematics Education Research Journal*, 29(3), pp. 349–367.

Ong, K. (2011). Reflection for action in the medical field, *Reflective Practice*, 12(1), pp. 145–149.

Pant, M. (2014). Participatory action research, in D. Coghlan and M. Brydon-Miller (Eds.). *The Sage encyclopedia of action research*, New Delhi: Sage, pp. 583–587.

Parrott, L. (2010). *Values and ethics in social work practice*, Exeter: Learning Matters Ltd.

Peeters, J. (2012). A comment on 'climate change: social workers' roles and contributions to policy debates and interventions', *International Journal of Social Welfare*, 21, pp. 105–107.

Popke, J., Curtis, S., and Gamble, D. (2014). A social justice framing of climate change discourse and policy: adaptation, resilience and vulnerability in a Jamaican agricultural landscape, *Geoforum*, 73, pp. 70–80.

Postholm, M.B., and Skrovset, S. (2013). The researcher reflecting on her own role during action research, *Educational Action Research*, 21(4), pp. 506–518.

Ranganathan, M., and Bratman, E. (2019). From urban resilience to abolitionist climate justice in Washington, DC, *Antipode*, pp. 1–23. Retrieved from https://doi.org/10.1111/anti.12555 [Last accessed on 10 November 2019].

Reed, S.O., Friend, R., Jarvie, J., Henceroth, J., Thinphanga, P., Singh, D., Tran, P., and Sutarto, R. (2015). Resilience projects as experiments: implementing climate change resilience in Asian cities, *Climate and Development*, 7(5), pp. 469–480.

Reitan, R., and Gibson, S. (2012). Climate change or social change? Environmental and leftist praxis and participatory action research, *Globalizations*, 9(3), pp. 395–410.

Reyes, J.A.C. (2016). *Disaster citizenship: survivors, solidarity, and power in the progressive era*, Chicago, IL: University of Illinois Press.

Ruch, G. (2002). From triangle to spiral: reflective practice in social work education, practice and research, *Social Work Education*, 21(2), pp. 199–216.

Santha, S.D. (2015). Early warning systems among the coastal fishing communities in Kerala: a governmentality perspective, *Indian Journal of Social Work*, 76(2), pp. 199–222.

Santha, S.D., Jaswal, S., Sasidevan, D., Khan, A., Datta, K., and Kuruvilla, A. (2016). Climate variability, livelihoods and social inequities: the vulnerability of migrant workers in Indian cities, *International Area Studies Review*, 19(1), pp. 76–89.

Santha, S.D., and Sunil, B. (2009). A malady amidst chaos: examining population vulnerability to the chikungunya epidemic in Kerala, India, *Loyola Journal of Social Sciences*, 23(2), pp. 111–129.

Schlosberg, D. (2012). Climate justice and capabilities: a framework for adaptation policy, *Ethics and International Affairs*, 26(4), pp. 445–461.

Schön, D.A. (1983). *The reflective practitioner: how professionals think in action*, New York: Basic Books.

Schön, D.A. (1985). *The design studio: an exploration of its traditions and potentials*, London: RIBA Publications.

Schön, D.A. (1987). *Educating the reflective practitioner: toward a new design for teaching and learning in the professions*, San Francisco, CA: Jossey-Bass.

Sinha, H. (2006). *People and forest: unfolding the participation mystique*, New Delhi: Concept Publishing Company.

Snow, D.A., Rochford, E.B., Jr., Worden, S.K., and Benford, R.D. (1986). Frame alignment processes, micromobilization, and movement participation, *American Sociological Review*, 51(4), pp. 464–481.

Southern, N. (2015). Framing inquiry: the art of engaging great questions, in G.R. Bushe and R.J. Marshak (Eds.). *Dialogic organisation development: the theory and practice of transformational change*, Oakland, CA: Berrett-Koehler Publishers, pp. 269–289.

Taylor, M. (2015). *The political ecology of climate change adaptation: livelihoods, agrarian change and the conflicts of development*, London: Routledge-Earthscan.

ter Mullen, R., Arts, W., and Muffels, R. (2001). Solidarity, health and social care in Europe: introduction to the volume, in R. ter Mullen, W. Arts, and R. Muffels (Eds.). *Solidarity in health and social care in Europe*, London: Springer-Science + Business Media, B.V., pp. 1–11.

Tester, F. (2013). Climate change as a human rights issue, in M. Gray, J. Coates, and T. Hetherington (Eds.). *Environmental social work*, London: Routledge, pp. 102–118.

Thompson, N., and Pascal, J. (2012). Developing critically reflective practice, *Reflective Practice*, 13(2), pp. 311–325.

Tierney, K. (2015). Resilience and the neoliberal project: discourses, critiques, practices – and Katrina, *American Behavioral Scientist*, 59(10), pp. 1327–1342.

Till, K.E. (2012). Wounded cities: memory-work and a place-based ethics of care, *Political Geography*, 31, pp. 3–14.

Tronto, J.C. (1993). *Moral boundaries: a political argument for an ethic of care*, London: Routledge.

UNDP. (2007). *Human development report 2007/2008 – fighting climate change: human solidarity in a divided world*, New York: Palgrave Macmillan.

Walters, C.J. (1986). *Adaptive management of renewable resources*, New York: Macmillan.

Ward, M. (2016). Rethinking social movements micromobilization: multi-stage theory and the role of social ties, *Current Sociology*, 64(6), pp. 853–874.

Whyte, K.P. (2019). Way beyond the lifeboat: an indigenous allegory of climate justice, in K.-K. Bhavnani, J. Foran, P.A. Kurian, and D. Munshi (Eds.). *Climate futures: reimagining global climate justice*, London: Zed Books, pp. 10–18.

Whyte, K.P., and Cuomo, C. (2017). Ethics of caring in environmental ethics: indigenous and feminist philosophies, in S.M. Gardiner and A. Thompson (Eds.). *The Oxford handbook of environmental ethics*, New York: Oxford University Press, pp. 234–247.

Williams, M.J. (2017). Care-full justice in the city, *Antipode*, 49(3), pp. 821–839.

Wisner, B., Blaikie, P., Cannon, T., and Davis, I. (2004). *At risk: natural hazards, people's vulnerability, and disasters*, 2nd edition, London: Routledge.

Young, I. (2003). From guilt to solidarity: sweatshops and political responsibility, *Dissent*, 50(2), pp. 39–44.

Zambo, D. (2014). Theory in the service of practice: theories in action research dissertations written by students in education doctorate programme, *Educational Action Research*, 22(4), pp. 505–517.

4   Ethical adaptation to climate change

Introduction

Philosophers consider climate change as fundamentally an ethical problem that has moral connotations, which cannot be neglected (Gardiner and Weisbach, 2016; Kronlid, 2014). Ethical thinking helps human beings to take the right decisions (Light and Rolston, 2003). As social workers, we are guided by certain basic values of service, social justice, dignity and worth of the person, importance of human relationships, integrity, competence, human rights and scientific inquiry. Other core values that guide our practice include empowerment, strengths, beneficence, non-maleficence, equality and harmony. The valuing of difference is yet another important characteristic that could guide our practice (Parrott, 2010). We invariably work with those sections of the society who form the most disadvantaged and oppressed groups in society; and whose conceptualisation of good life might be quite different from that of other mainstream groups. Their marginality, powerlessness and vulnerability contexts can be the starting point of our practice, and it requires an empathetic understanding of their lived experiences (ibid).

COMEST (2013) has identified some of the key principles that could guide adaptation such as avoiding harm, ensuring justice, solidarity, fairness and equitable access to resources. Certain ontological questions that emerge in the context of climate change adaptation are: Adaptation to what? Who is to benefit from the adaptive innovation? And how? Climate change adaptation raises important ethical concerns such as who gains and who benefits from adaptation strategies and the legitimacy of such responses (Adger et al., 2017). All adaptation decisions have justice implications (Adger et al., 2006). And these decisions and the actions undertaken can have intersectional, intergenerational and interspecies implications. Intersectional contexts refer to the situatedness among diverse community actors in terms of their historicity and social positions or locations and can be understood in terms of their politics of differences. These differences could be in the form of their identities represented in the form of gender, race, caste, ethnicity, disability, employment and so on. Intergenerational implications refer to the consequences of our present decisions and actions on the future generations as well. Moreover, it insists towards the blending of empathy and knowledge between the older and younger generations of the present. An

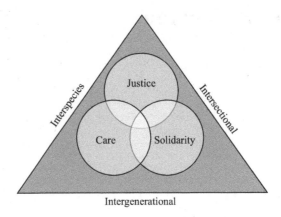

Figure 4.1 Ethical adaptation to climate change.

ethical adaptation to climate change is not just about the humans. It is about the interconnectedness of life and its systemic organisation, and therefore our decisions and actions should also take into account the balanced existence of all humans and non-humans in this earth. The interspecies aspect of adaptation is therefore equally relevant when we speak about ethical adaptation. It is in these contexts of intersectional, intergenerational and interspecies dimensions of adaptation that the adaptive innovation model envisages the presence of three critical elements as crucial for ethical adaptation. These are *climate justice*, *ethics of care* and *solidarity* (refer to Figure 4.1).

These key elements of ethical practice that could shape adaptation projects are elaborated in detail as follows.

Climate justice

The threats posed by climate change and extreme hazard events raise important social and ecological justice questions. Most of the climate justice issues are not only an environmental phenomenon, but also a social, economic and political issue (Hetherington and Boddy, 2013; Joseph, 2017; Peeters, 2012; Reitan and Gibson, 2012; Santha et al., 2016). In this regard, the climate justice framework argues that the most marginalised and subjugated actors should get privileged access to adaptation such that they have sufficient adaptive capacity to pursue a decent life (Grasso and Feola, 2012). An ethic of justice focuses on questions of 'fairness, equality, individual rights, abstract principles, and the consistent application of them' (Held, 2006, p. 15). It seeks a fair solution between competing individual interests and rights, and it strives to protect equality and freedom (ibid). Rawls (1971) in his Theory of Justice has identified distributive and procedural elements of justice, where he prioritises human rights over public goods; and insists that society should be governed by principles of inclusive governance that seek to protect and enhance the quality of the poorest (Pelling, 2011). Distributive justice refers to the distribution of the beneficial and adverse effects of climate change and adaptation, while procedural justice refers to the degree of recognition and participation (Adger et al., 2006).

In South Asia and African regions, climate change impacts are largely felt by vulnerable groups such as small-scale farmers, landless labourers, artisanal fisherfolk, migrant workers in the informal sector, single women, children and elderly people from marginalised communities, slum dwellers and homeless population due to their heightened exposure and limited adaptive capacities (Kwan and Walsh, 2015; Mearns and Norton, 2010; Santha et al., 2016). The distributive justice framework could help us to identify the nature of distribution of resources that could enhance human capabilities among vulnerable individuals and groups (Kronlid, 2014). It pertains to the fairness in the distribution of climate change burdens and benefits (Page, 2007). It is widely believed that providing these vulnerable groups with adequate resources would enable them to adapt effectively to the impacts of climate change (Peeters, 2012). Prominent approaches to enhance adaptive capacities of vulnerable groups include programmes that transfer income or assets to the poor, social insurance, social assistance and labour market regulations (Barrientos, 2010; Devereux and Sabates-Wheeler, 2004; ILO, 2001; Leichenko and Silva, 2014).

S.F.F.4.1 The vulnerability contexts of nomadic pastoralist communities.

Last year, I met a team of researchers from Kenya and Ethiopia. They were involved in understanding the livelihood struggles of nomadic pastoralists in these countries. There are different tribes of nomadic pastoralists for whom their livestock represent a significant component of their culture, livelihood and lifeworld. Severe drought situations have forced many among the nomadic pastoralist communities to reduce their ownership of livestock. Factors such as severe water scarcity and soil erosion have worsened their livelihood struggles. Highlighting their extreme vulnerability, Ferdinand (2019) notes that some among these communities had to depend on international food aid for their everyday survival. However, he observes that the herders have begun to give up cattle in favour of camels and terms it as a kind of transformative adaptation (ibid). One could witness similar livelihood struggles among the nomadic pastoralist communities located along the Rann of Kutch in India. Since the last few years, these communities are severely affected by extreme drought situations. There has been severe scarcity of both water and fodder. Their adaptation strategy to migrate to other places were also constrained due to diverse factors such as resource scarcity, urbanisation and fear of conflicts. Disruptions in the traditional resource sharing arrangements that they earlier had with host farming communities in their regular migration routes have also affected their livelihood practices. And many among them, such as people from the 'Maldhari' community had to end up in distress sale of their livestock (Ravi and Vanak, 2019). These communities also have very limited access to social protection and drought relief programmes. In some regions, the government had opened up cattle camps to help people to cope with the drought and fodder crisis. However, very few people belonging to the nomadic pastoralist communities are able to derive benefits out of such schemes. While the district administration offers drought relief to farmers and semi-pastoralists who own cattle and buffalo by providing them fodder subsidies and access to fodder depots, the nomadic pastoralists who own sheep, goat and camel often do not get such benefits. Moreover, the subsidies cater to the needs of only those people who own some land and not for the largely landless communities such as the nomadic pastoralists.

In practice, these strategies of distributive justice may not be adequate to ensure effective and just adaptation (Bulkeley et al., 2013, 2014). There are ethical limits to adaptation if the consequences of adaptation actions could threaten peoples' well-being (Adger et al., 2006). Critiques point out that adaptation projects today necessarily do not account for equity or about creating system transformations. Uncapher and Yvellez (2019) cite examples of constructing sea walls as solutions that are blind to both equity and system resilience. Moreover, such adaptation strategies can exaggerate existing inequalities or generate new ones (Kronlid, 2014; Pelling, 2011). The implementation of distributive justice that is largely governed by techno-centric development interventions and mediated by local elites and outside experts could reinforce the vulnerabilities of poor and marginalised groups (Klepp and Chavez-Rodriguez, 2018). Critiques also point out that the main beneficiaries of present-day adaptation strategies are private contractors, consultants, architects, designers and corporates, and necessarily not the vulnerable groups (Hoggett et al., 2009; Ranganathan and Bratman, 2019; Uncapher and Yvellez, 2019). On many occasions, adaptation measures are only aimed at protecting at-risk assets in high-value areas rather than building the adaptive capacities of vulnerable populations around high risk areas; and in such contexts, climate change can be a threat multiplier for these vulnerable groups (Uncapher and Yvellez, 2019).

In contrast, there are other perspectives that look at adaptation from procedural and structural justice lens (Adger et al., 2006). Procedural justice approaches are concerned with the issues of access to decision-making structures, governance of resources and related adaptation practices (Adger, 2013). It largely corresponds to fairness in access to democratic decision-making by diverse actors (Adger et al., 2006; Young, 1990); and is largely shaped by the issues of representation, voice and the distribution of power in decision-making (Adger, 2013). It is also focused towards strengthening the deliberative and participatory procedural principles in adaptation (Kronlid, 2014). The notion of structural justice examines how prevalent social structures including social positions and hierarchies, rules and norms pertaining to resource access and use create inequities, social exclusion and injustices through adaptation practices (ibid). Such a lens could throw more insights on how vulnerabilities are constituted in and through social structures; and how certain adaptation practices could worsen social inequalities and result in maladaptation. Adaptation has to be thus associated with a deeper understanding of societal change rather than mere adjustment to a crisis. It is in this context that many climate activists and scholars argue for a practice framework that strives towards the provision of rights, responsibilities and recognition (Bulkeley et al., 2014; Robinson, 2019). Adaptive innovation has to be a socially inclusive and people-centred process, where outcomes are in itself the recognition, representation and empowerment of the most marginalised and subjugated community actors. If founded on these inclusive principles of recognition and empowerment, climate change adaptation in itself can enable marginalised groups to shape decision-making and become owners of their own development pathways (Morchain, 2018).

Climate change adaptation decisions have justice consequences that are both intersectional as well as intergenerational (Adger et al., 2009). The processes of infusing climate justice cannot be isolated from the broader social and intersectional, historical, political and economic processes that impact upon vulnerable and marginalised groups (Moellendorf, 2012; Ranganathan and Bratman, 2019; Schlosberg and Collins, 2014). This is important because in many communities, harm associated with climate change is compounded by other factors rooted in history and similar development contexts (Whyte and Cuomo, 2017). The lack of recognition such as the denial of love, respect and esteem could result in different forms of hurt that are characterised by humiliation, disrespect and denigration (Honneth, 1995). Adaptation ensuring climate justice should therefore aim at addressing people's vulnerabilities and injustices that are rooted in the forces of history such as colonialism alongside capitalism and industrialisation (Whyte, 2019). At the same time, we need to be aware of those conditions that presently prevent vulnerable and marginalised groups to participate and shape fair adaptation strategies.

S.F.F.4.2 Structural inequalities and climate justice.

Post the deluge that devastated the state of Uttarakhand in India in 2013, I was involved in helping some of my former students in designing the livelihood recovery of few affected communities. During our interactions with flood-affected communities, we realised that the marginal farmers and landless households in these regions were severely affected by climate change. For instance, traditional farming practices were severely constrained by prevalent socio-economic inequalities. In my opinion, both gender and caste inequalities in these villages had worsened the vulnerabilities of populations to floods and landslide; and these disasters further reinforced the structural inequalities. We realised that caste hierarchies played an important role in constraining the adaptive capacities of marginal farmers. Farmers belonging to the less-privileged caste, and women amongst them in particular were equipped with very limited social capital and other livelihood assets to deal with the crisis; a crisis that is exacerbated due to a complex web of caste and gender inequalities, poverty and unequal power relations in the communities. These factors also prevented the less-privileged groups to access water sources during times of water scarcity. We also came across instances where widowed women who belonged to less-privileged castes were denied basic support services such as food, clothing and shelter post the floods.

The notion of power is central to adaptation (Pelling, 2011). The power held by each actor would shape the decision-making processes and outcomes of adaptation (ibid). Fair adaptation is also concerned about how decisions are made, and the issues of community, identity and ownership of the adaptation process gains prominence (Adger, 2013). We have to therefore develop an intuitive understanding of justice from grounded experiences and practices in responding to injustice (Barnett, 2011; Hobson, 2006; Williams, 2017). Drawing insights from the premises laid out by Amartya Sen (2009), a working definition of climate justice could incorporate readily available understandings of injustice, which is rooted in actors' judgements and intuitions of injustice and indignation (Barnett, 2011). In this regard, as discussed below, the ethics of care has the capacity to transform social relations in progressive ways and nurture new ways of collaboration among diverse actors (Conradson, 2011).

Ethics of care

Care is conceived as both a cluster of values and practices (Held, 2006). Care can act as a moral value and as a basis for the political achievement of a good society (Tronto, 1993). It emphasises on those context-specific and situated knowledge systems and practices that would nurture beneficial caring relationships as well as address histories of harm and injustice (Whyte and Cuomo, 2017). Public performance of care ethics such as in climate change adaptation would help in understanding the resistance of marginalised and subjugated actors to the denial of their basic rights and needs by outside neoliberal forces, which are interested only in profit maximisation and maintaining certain kinds of power relations (ibid). There are multiple perspectives that have contributed to the discourses pertaining to the ethics of care. In the context of climate change adaptation, two perspectives gather significance. These are the indigenous and the feminist perspectives of care.

The indigenous perspectives of care emphasise on the importance of awareness of one's place in specific social–ecological systems that believe in healing wounds of oppression and injustice through rebuilding and restoring relationships of inter-dependence and reciprocal responsibilities, and which are nurtured and supported by indigenous knowledge systems and practices (Whyte and Cuomo, 2017). Caring for a place or a particular social–ecological component is also a way to repair our worlds to create socially sustainable spaces (Till, 2012). According to the Kari-Oca 2 declaration, caring and sharing among other values are crucial in bringing about a more just, equitable and sustainable world (Indigenous Peoples of Mother Earth, 2012). Caring among indigenous communities motivates responsibilities involving reciprocity, harmony, solidarity and collectivity, which is also a foundational value for justice and sustainability (Whyte and Cuomo, 2017). The feminist care ethics emphasise on understanding individuals as embedded and interdependent actors whose mutually beneficially caring relationships are foundational for ethical values and practice with women and other intersectional and subjugated actors (ibid).

The ethics of care drawn from both indigenous and feminist perspectives emphasise on deep connections and moral commitments between humans and their social–ecological system, which could also guide ethical forms of environmental decision-making. While indigenous perspectives value the attentive caring to the intertwined needs of humans and non-humans in their social–ecological system, feminist perspectives give importance to the empowerment of communities to care for themselves and their social–ecological system. Moreover, a feminist ethic of care does point towards reconciliation between care and justice and an ethic in which all persons should have the right to care for and be cared for by others (Parrott, 2010). Climate change adaptation can be guided by the values and insights drawn from relation-centred traditions and practices of caring the self and other elements in the social–ecological system. Care, as both value and practice, is associated with all aspects of life, and not just the intimate and familial (Sevenhuijsen, 1998). Care as a practice is a complex phenomenon that involves both thought and action that are directed towards the same end (Tronto, 1993). However, the construction and practice of care varies from one culture to another or from one intersectional group to another (ibid). There is a need for a situational ethics of care, which represents an ethic that is rooted in the everyday decisions and practices of what is good, fair and care-full (Smith, 2009).

The ethics of care and justice are usually positioned as a binary. While justice is viewed more as a universal, rational, exterior to self and more appropriate for moral decision-making, the notion of care is often looked upon as an emotional, specific and embodied ethic (Williams, 2017). Such a binary positioning has placed justice in the public sphere, while restricting care to the private sphere (ibid). However, such a conceptualisation is problematic, specifically in the context of complex problems such as climate change adaptation. In contrast, the ethic of care has to be construed as a transformative ethic, which is more suitable for just, inclusive and creative adaptation (Till, 2012; Williams, 2017). For instance, we could locate the processes of everyday caring encounters embedded in the adaptation processes, where people interact and take care of their natural environment and this in itself could facilitate justice at the ground level (Fincher and Iveson, 2008). There are also other studies that have attempted to draw the linkage between the two perspectives of justice and care (Gould, 2008; Till, 2012; Williams, 2017). Some scholars have visualised justice as always in tandem with care (Ruddick, 1995). In contrast, there are studies, which argue that both justice and care have to be seen as distinct, but as interdependent and situated practices that could facilitate collaboration and co-responsibility among involved actors (Williams, 2017). Ethical adaptation to climate change needs to include insights from both ethics of care and ethics of justice. Nevertheless, too much integration could erode the unique values that each of these possesses (Held, 2006). Justice practices can also be articulated through a subtle ethics of care rooted in the daily practices and routines of the vulnerable population who are engaged with multiple complex issues and not just climate change (Parrott, 2010; Ranganathan and Bratman, 2019; Williams, 2017). These practices may overlap with other arenas such as food and livelihood insecurity, lack of transit, homelessness and gendered burdens (Ranganathan and Bratman, 2019).

Care as a practice enables us to build trust and mutual concern and connectedness between persons (Held, 2006). Moreover, relations of trust serve as the most important personal and social asset (ibid). It is these reciprocal relations that the values of caring are exemplified. In her own words, Held (2006, p. 42) says,

> To work well, societies need to cultivate trust between citizens and between citizens and governments; to achieve whatever improvements of which societies are capable, the cooperation that trust makes possible is needed. Care is not the same thing as trust, but caring relations should be characterised by trust, and caring and trust sustain each other.

An ethics of care perspective would help us to look at climate change adaptation as a practice towards creating worthy affiliations and relationships and at the same time striving towards healing the earth and its citizens. It is not only the caring between the humans that matters, but also caring the social–ecological environment in which humans and non-humans survive. The Chipko movement in India, for example, is an environmental movement that was grounded in ethics of caring and responsive care taking (Whyte and Cuomo, 2017). In such movements, the relationships and contexts drive the moral value of human rights

(Carmalt, 2011). This implies that only the relationships in a given context can determine which rights are more or less valuable. The lived and situated experiences of community actors and the relationships in which their actions are embedded could therefore play an important role in the recognition and practice of certain rights (ibid).

Care has to be looked upon as not something special but that which forms a normal part of every-day life in which actors engage and depend upon (Parrott, 2010; Williams, 2016). We need to explore the localised articulation of care that are often expressed in their own unique verbal and non-verbal languages of community actors. In this regard, there is a need to re-humanise climate change and its impact on the everyday practices, spaces and routines of vulnerable populations (Ranganathan, 2017). Moreover, justice in everyday life is not static, instead is always a dynamic, situated, relational and negotiated practice (Williams, 2016). The co-designing of adaptation strategies should therefore be based on the everyday, ordinary, mediated and direct practices of care and justice (Williams, 2017: 826). And it could be practiced more in terms of a messy, relational, imperfect and ongoing ways, and yet are capable of healing and repairing our world (Till, 2012; Williams, 2017). It is also important to be sensitive to the needs of different vulnerable groups, the dynamics of their interactions and to facilitate their involvement accordingly. Appreciating different standpoints, being compassionate, demonstrating thoughtfulness, listening empathically to disagreements and struggles could all be functions of this role.

In the context of problems such as global environment change, 'the ethics of care offers a view of both the more immediate and the more distant human relations on which satisfactory societies can be built' (Held, 2006, p. 28). As social workers or as care givers, we not only have the appropriate motivations in providing care but are also skilled in providing effective practices of care (Held, 2006). The practice of care requires a deep and thoughtful understanding of the situation, and of all the actors' situations, needs and competencies (Tronto, 1993). Caring also requires a variety of resources (ibid). During our caring encounters, we do not seek primarily to further our own individual or organisational interests. Instead, our interests are intertwined with those of the actors whom we care for (Held, 2006). Caring, thus, is a value-embedded practice that seeks to preserve and promote human relations between us and particular others (ibid). It involves the cooperative well-being of those in the relation and the well-being of the relation itself (ibid). We have to be extremely cautious that our interventions do not normalise the language of climate change and impacts or maintain the status quo, which assumes that vulnerable populations would have endless capacity to adapt and their marginalisation processes are normal and acceptable (Ranganathan and Bratman, 2019). The ethical principle of care or the duty to care is thus very important when working with vulnerable populations.

The four values of care, namely caring about, taking care of, care-giving and care-receiving could be understood as analytically separate, but interconnected phases (Tronto, 1993). 'Caring about' asks actors to recognise that care is necessary and remain attentive to the need in others. The situational analysis and micro-mobilisation phases in the adaptive innovation model could address certain crucial elements of caring about. 'Taking care of' implies the capability and responsibility of actors to implement the realisation from caring about into action. It involves assuming responsibility towards the identified need and determining how to respond to (ibid). The dialogic ideation and action framing phase could represent the acts and spaces of taking care of. 'Care giving' is a feeling of responsibility for doing the work of caring. It involves the direct meeting of needs of care (ibid). The very act of caring also produces certain elements of competence in caring work. The piloting phase and the implementation practices in adaptive innovation could represent certain features of giving care and receiving care. 'Receiving care' ensures that the care work has been done and it has made things better in a responsive, inclusive and open manner (ibid). This phase recognises that the actors receiving care will respond to the care she or he receives (ibid). However, in complex problems such as climate change, receiving care cannot be understood or felt in a linear simplistic manner and is the most difficult of the four types. This characterises the emergence phase of adaptive innovation. We, as social workers have to engage from the standpoint of the other, keep listening to others, and at the same time enhance our own sense of self-worth (Till, 2012). Good care requires that all these phases and ethical elements of care fit together into an appropriate whole (Tronto, 1993). Community actors involved in climate change adaptation has to believe that their participation is beneficial to themselves and there is a legitimate space for dialogue. In this regard, we have to equip ourselves with the mind-set and skills to work with real issues that infuse community actors with vibrant and energetic forms of climate action (Averbuch, 2015). A crucial role would be to help foster or accelerate new ways of talking, thinking and doing that could lead to meaningful social and ecological change. This requires mindfulness, humility and authenticity on the part of the social worker. We also need to maintain the right attitude of being patient and culturally sensitive, open-minded and empathetic to the day-to-day lived realities of community actors.

Solidarity

Solidarity is both a political as well as ethical goal (Mohanty, 2003). It entails the recognition that liberation is a collective project (Freire, 1970). Maintaining a commitment to confronting oppression in all forms is a key mandate of solidarity (Jobin-Leeds and AgitArte, 2016). It thus propagates a collective stand against structural injustice such as oppression, exclusion and exploitation (Naidoo, 2017). In modern complex societies, one has to view solidarity in the context where every actor is free from being collectively denigrated and is not marred by experiences of disrespect (Honneth, 1995). Vulnerable and dependent people will count on solidarity, which is a reciprocal support at both expected and unexpected moments (Sevenhuijsen, 1998). In the context of collective action and social movements, political solidarity could be understood as a conscious commitment by a group of community actors to challenge a situation of injustice, oppression, tyranny or social vulnerability (Scholz, 2008). In a similar vein, feminist and critical theorists view solidarity as 'political action on collective terms' (Kolers, 2016, p. 5).

Solidarity has an important role towards sustaining both care and justice. Though solidarity was cast as something typically to build justice in a welfare state, such an approach did not look at the challenges posed by the diversity and inequality among the citizens (Kolers, 2016). Solidarity therefore was looked upon as a means to achieve distributive justice (ibid). In contrast, Sevenhuijsen (1998) argues that such an approach encourages a solidarity of giving and taking, which projects a norm of sameness and uniform normality. Instead, we need to develop diverse forms of solidarity that has scope to explore differences and situatedness of actors' situations in their everyday struggles to live with dignity. The notion of solidarity thus derived would possess a political meaning to care and mutual commitment. Solidarity without care (which is largely propagated through the distributive justice paradigm of sameness) results in an impoverished sense of morality and collective responsibility. It also has the potential to strengthen the privatisation and moralisation of care (ibid). On the other hand, there is a need for caring solidarity that could recognise the diversity of needs and provide collective support accordingly.

Solidarity in climate action primarily rests on the recognition of both difference and equality (Blau, 2017; Olthuis, 2000). This necessitates for us to identify a more complex and internally differentiated conception of solidarity, which tolerates individual differences and is compatible with pluralism (Alexander, 2006). This is because actors in a specific social–ecological system are 'symbolically represented, as independent and self-motivating individuals responsible for their own actions who feel themselves, at the same time, bound by collective solidarity to every other member of this sphere' (Alexander, 2006, p. 402). It is the commonalities of experiences, histories and intersectional identities that form the basis for solidarity towards organising marginalised and subjugated actors (Mohanty, 2003). It demands a space for radical equality that recognises differences with dignity, where all vulnerable populations receive special protections (Blau, 2017; Olthuis, 2000). Forging solidarity thus requires action with empathy (Doherty and Doyle, 2013). It also requires working across a whole range of differences and intersectionalities (von Kotze and Walters, 2017; Young, 1990). This implies that we have the responsibility to establish caring and committed relationships that preserve community autonomy and recognise the knowledge systems embedded in the local culture (Doherty and Doyle, 2013). It also necessitates that those most affected are facilitated to lead and speak for themselves by using new images and languages that would disrupt oppressive thinking and practices (Jobin-Leeds and AgitArte, 2016).

A multiculturalist approach that legitimises the identities, differences and situatedness of marginalised and subjugated actors is very crucial in this regard (Alexander, 2006). Multiculturalism implies that marginalised and subjugated actors are accepted because their qualities are also accepted (Alexander, 2006; Kivisto and Sciortino, 2015). It is thus an ethical value that could stimulate acceptance and recognition of particular differences among diverse actors in a social–ecological system. We have to ensure that a diversity of experiences and ideas are heard, and our attention has to be on the multiplicity of social locations and identities of marginalised actors and how these social locations are linked to the historical and structural forms of domination (Walters and Butterwick, 2017). In this regard, it is important that the adaptation projects address a shared concern and not the agenda of a few dominant players in the community or outside. Such an understanding towards forging solidarity suggests 'great respect for individual capacities and, at the same time, trust in the goodwill of others' (Alexander, 2006, p. 403). It involves overcoming differences by creating bonds of cooperation (Blau, 2017).

Solidarity calls for human relationships with a sense of obligations and a set of actions (von Kotze and Walters, 2017). It is grounded in the common values and shared moral commitments of the community (Oosterlynck et al., 2017). Mutual interdependence is a feature of solidarity (de Beer and Koster, 2009). Certain values integral to solidarity in climate change adaptation are empathy, cooperation and collaboration (Blau, 2017). Mohanty (2003) identifies the central values shaping solidarity in terms of mutuality, accountability and the recognition of common interests amidst diversity and difference. Solidarity combined with reflexivity and commitment to participative social behaviour could be viewed upon as the social and cultural base for justice (Arts et al., 2001). In our everyday practice, it can also be understood as a community of trust, which is characterised by diverse forms of mutual obligations (Bayertz, 1999; Krol, 2016). It invokes community actors to identify themselves with the community, sustain emotional ties and practices of mutual help and maintain necessary public legitimacy (Bayertz, 1999). It also entails positive collective moral obligations and responsible practices (von Kotze and Walters, 2017). Climate change adaptation has to be pursued as a collective and responsible practice that articulates justice through social connections with shared responsibility across different spatial relations of power, privilege, interest and capacity for action (Blau, 2017; Young, 2004).

The relationships of solidarity are transitive, deliberate and indicate a purposeful commitment to interdependence and reciprocity as values and outcomes (von Kotze and Walters, 2017). These relationships at its core are asymmetrical, mutual and reciprocal in nature (Grieves and Clark, 2015). It is often expressed as a feeling or an attitude embedded in action, and in itself can be understood as a desirable situation (de Beer and Koster, 2009). Building meaningful partnerships based on trust and reciprocity is a key strategy towards forging solidarity. We need to build partnerships striving towards sustainability that begins at the local level and seeks ultimately to affect the wider world. And such an approach requires trust, open-mindedness, complete fairness and lots of energy (Blau, 2017). Yet experience suggests that those practitioners who are able to empathise with the target communities and communicate in ways that resonate with those populations forge meaningful partnerships embedded in solidarity. There has to be a relational ethic that recognises and values mutual respect, dignity and connectedness between us and the participant actors. On many occasions, this also involves working with people at different levels in carefully planned, cross-disciplinary manner who are able to delve deeply into the meaning of adaptation and well-being. Forging solidarity involves organising and mobilising, which is a process of collective visioning, action and reflection. This invokes an active commitment and deliberative engagement from all concerned actors to empower subjugated actors to meet their care needs in best possible ways. It expresses a commitment to sustain trusted and reciprocal caring arrangements during everyday risk situations and complex emergencies as well.

Solidarity is a process of stepping forward, standing with and staying connected (Walters and Butterwick, 2017). These processes are part of everyday relationships, moment-to-moment engagements and both successes and failures. The whole interventions call for ongoing engagement, humility and openness as there can be many ethical grey areas. Intersectional variations in solidarity can be observed for example, when male members value central authority, hierarchy, competition, analytical reasoning at the cost of what women value in some societies such as mutual trust, equality, cooperation, intuition, empathy and interpersonal skills (Hirota, 2009). Resource scarcity, changes in family and community structure, retreat of the welfare state and unregulated penetration of the market into the social worlds of community actors can negatively affect intersectional and intergenerational solidarity in a specific locale (ter Mullen et al., 2001). On many occasions, solidarity-based processes could replace the centralised and hierarchical state institutional structures and functions that implement planned adaptation strategies (Reyes, 2016). However, in contrast, everyday solidarities that are often informal, when formalised and made hierarchical can also turnout to be less efficient and the acceptance levels can also come down (ibid).

Strengthening solidarity requires a careful forging of links between individuals, the personal and collectives (Vielmas, 2017). Collective action can be nurtured only if individual experiences and associated feelings are analysed and collectively processed. Each actor has to be considered as an active agent of change. Each individual actor should be recognised as capable of contributing to the process of knowledge generation; and also have the right to play an active role in shaping interventions that affect their own well-being and that of other community actors. Such strategies of forging solidarity require meaningful dialogue, where listening to the rationalities of diverse actors is an essential skill (ibid). In this regard, all participant actors should have the opportunity to accept or decline from participating in the adaptive innovation process. Guided by this principle, the community review of the proposal to initiate the adaptive innovation project is a crucial step towards getting the initial informed consent. Forging solidarity also requires the creation and elaboration of a narrative, which makes it possible for other actors to identify and reflect with one's own circumstances (Alexander, 2006; Kivisto and Sciortino, 2015). This requires a facilitation of more and regular human contact and interactions that is blended with cultural creativity and political competence (Alexander, 2006; Cruz-Saco and Zelenev, 2010; Kivisto and Sciortino, 2015).

Face-to-face interactions and group meetings are very important to nurture motivation and build solidarity. Face-to-face interactions enable actors to filter vast amounts of information and accordingly orient themselves for action (Clark, 2009). It has been found that conflicts are resolved through the face-to-face interactions among diverse actors, as these meetings create a familiar space for these actors to negotiate and arrive at meaningful solutions (Doherty and Doyle, 2013). Cultivating solidarity requires sensitive explorations of experiences and interpretations (Walters and Butterwick, 2017). The lived experiences and the nature of place, context and identity shapes solidarity among actors (Doherty and Doyle, 2013). Community actors do not always engage in adaptation strategies through relationships of power alone. As part of their everyday lived experiences, what matters more importantly is their feelings for others, and which are largely structured by the boundaries of solidarity (Alexander, 2006). In this regard, storytelling enables us to build affective solidarity, which gives adequate attention to actors' emotions and lived experiences (Walters and Butterwick, 2017). Discourses in the form of narratives and stories that generate dialogue enable community actors to articulate and legitimise both their actions and the situations that give rise to them. Nevertheless, care has to be taken that stories are not misinterpreted within unequal structures of power that exists between the storytellers and the listeners (Glass and Newman, 2015). Storytelling and listening therefore requires critical self-reflexivity (Walters and Butterwick, 2017). Nurturing solidarity, thus, requires being open to 'the other' by being aware of how our own perspective and social location shapes our worldview and ignorance as well (Masschelein, 2010).

Forging solidarity among diverse actors is a challenge. The forces of modernisation today have changed the nature of trust and solidarity (Beck, 1992; Giddens, 1994). There has been considerable erosion of trust among local relationships; and therefore, we have to engage in building active trust among diverse local actors. Active trust that is reflexively built through 'a mutual process of disclosure' could be the foundation to build new forms of solidarity (Giddens, 1992, p. 121). One has to forge solidarity, which requires every actor 'to be vulnerable, to trust, to love, to hope' (von Kotze and Walters, 2017, p. 5). Solidarity can be 'built in a slow, careful, step-by-step sustained process in which relationships are nurtured with care, and a commitment to mutuality is part of the process' (von Kotze and Walters, 2017, p. 8). This also involves renewal of personal and social responsibilities, stimulating political action through the agency and involvement of community actors, risk reduction through empowering community actors and providing means for social inclusion and participation (Giddens, 1992, 1994). In this context, it has become very crucial that we give attention to the everyday forms of solidarity, so as to examine how ordinary day-to-day social encounters can also result in collective action and risk reduction (Reyes, 2016). During disaster situations, people rely on their everyday networks and practices of solidarity that has scope to pay more attention to shared feelings and symbolic commitments that actors have towards one another (Alexander, 2006; Reyes, 2016).

The focus on participation of poor and marginalised groups does not entitle to shift the entire burden of adaptation on them. Care also has to be taken that the emphasis on participation does not raise unrealistic expectations of local community actors. Participation does not mean that outsider participation will not be there. The design of adaptive innovation is such that decision-making also involves outsiders and experts who would co-participate with local community actors to identify and address adaptation issues. However, this requires continuous sensitivity on how power relations and knowledge interfaces are shaping the partnership and continuous inquiry into the process of collaboration as well. Care should be taken that local communities are not overpowered by outside ideas and processes. Community actors need to be viewed as experts in lived experience and grounded knowledge of their oppression and survival. On the other hand, outsider experts may offer technical skills and other spheres of knowledge and expertise. In this regard, no one way of knowing has to be valued over another, and possibilities of blending diverse knowledge systems have to be explored (Agrawal, 1995). An effective and efficient process of information sharing and communication between diverse institutional actors therefore need to be in place, taking into account the diversity in knowledge, skills and interests of different community actors. Our work with the innovation platforms has to therefore aim at bringing together different sources of knowledge and expertise and thus develop new community-based solutions. We, therefore, have an important facilitating and enabling role in the adaptive innovation processes. We need to be knowledgeable of the concerned climate risk situation that is being addressed and should have the professional skills to bring diverse stakeholders together to develop ethical adaptation strategies.

S.F.F.4.3 Care and solidarity among traditional riverine fisherfolk.

Few years ago, I collaborated with a few traditional riverine fishing communities of Kerala in South India to understand their cultural components, local knowledge systems and collective action strategies. I discovered that their fishing practices were based on profound relationships of care and solidarity. Their notion of care was inherent in their understanding of the interlinked ecosystems and its relationship to the larger social–ecological system. For instance, they widely believe that the modernisation and commercialisation of paddy farming along the river basin has considerably depleted the soil and fishery resources in the region. They narrated that the shift from the cultivation of indigenous varieties of rice to hybrid varieties, and a huge increase in use of chemical fertilizers and pesticides have destroyed the breeding and nursing grounds of fishes. Recounting such instances, traditional fisherfolk emphasise on the need to maintain a 'careful' synergy between the various eco-systems such as the forests uphill, the paddy fields and surrounding wetlands, the rivers and the backwaters flowing towards the ocean. Guided by such belief and value systems, traditional fishermen also have evolved the practice of caring and responsibly managing the fisheries.

I observed that these fishing communities never encourage the practice of capturing fries or juveniles. According to them, 'capturing young fries or juveniles of fish varieties is a matter of extreme dishonour that could bring disrepute to their identities of being a "true" or "expert" fisherman'. Therefore, these fishermen leave the fries and juveniles that are trapped in their nets back to water. In a similar vein, they also abstain from 'unethical' fishing practices such as using chemicals, herbal poisons, explosives and nets with very small mesh size. Their norms of resource allocation through which they maintain the practice of allocating fishing rights over particular fishing sites also showcase elements of care and responsible fishing practices. According to the norms, 'a fisherman or his group gains the right to fish from a particular fishing site if he reaches the fishing site first, ahead of other fishermen'. Such a norm is also backed by a strong belief system that violating this norm would result in poor or no yield from that site in the future. Such norms and practices prevent the over-pressurising of rich fishing sites and exploitation of specific fish varieties. The traditional fishermen also have the practice of shifting from one fishery to another with the change of seasons. The practice of shifting within fisheries limits the exploitation of a targeted stock, as fish are captured only during their peak availability and are not driven by market demand. I also observed that these fishing communities practice fishing holidays during certain seasons and most often these holidays coincide with the spawning and breeding time of many riverine fish varieties.

These caring norms and practices are reinforced and recursively constituted through their solidarity networks. These 'caring solidarity' networks are located at various levels, be it within fishing groups, their trade union and in their interactions with other marginalised groups such as landless farmers and labourers from less-privileged castes. These networks nurture and sustain reciprocal relationships based on common interests, trust and commitments valuing the health and sustainability of fragile social–ecological systems. These solidarity networks also engage as pressure groups to shape government policies and decisions pertaining to their demands of justice. However, in my opinion, the State and other policy makers consider the demands of fisherfolk only from the perspective of distributive justice. Consequently, the welfare measures of the State have created a dependency relationship, where the State is always considered as the 'provider' while the fishing communities are looked upon as the 'receiver' of aid.

Further, climate change complimented by neoliberal adaptation designs have placed the livelihood security of these traditional fishing communities in peril. For instance, today aquaculture and aquaponics is being propagated by the State and the market as a successful adaptation measure for paddy farmers. Aquaculture has already proved its potential to damage fragile and virgin ecosystems (Shiva, 2000). Aquaponics is an entrepreneurial farming strategy that integrates the cultivation of vegetables with fish. Many traditional riverine fisherfolk will not be able to adopt such designs, as they neither do have the land nor other capital resources to start such a venture. The investment costs are also very high that both poor farmers and fisherfolk will not be able to afford. In addition, according to the fisherfolk, 'the present model uses water hyacinth as the base, which is a dangerous exotic weed that has already affected the health of their wetland ecosystems'. In contrast to such designs, they believe that the adaptation designs should be 'linked more closely to their social-ecological system and everyday livelihood practices'. According to them, adaptation embedded in justice includes 'strengthening the ecological health of the resource system by recognising their values and knowledge systems and ensuring that their fishing rights are always maintained and upheld'. However, they observe with concern that the present pathways of neoliberal development accompanied by the tokenistic nature of welfare provisioning have begun to exclude these communities from their customary common property resources. Their only hope is now on their own solidarity networks and collective action.

Our crucial role in adaptive innovation is to co-create knowledge, improve practice and transform the lives of participant actors. Building the adaptive capacities of participant actors and thereby strengthening the pathways of climate justice is the central focus of adaptive innovation. In this regard, knowledge is inextricably linked to practices shaping both the actors' everyday lives as well as imparting broader systemic changes. However, behavioural change and social transformation in adaptation will take time to sink in and is sometimes hard to measure in terms of impacts. Donors and development practitioners have to be ready to accommodate these non-linear time spans and uncertain, non-quantifiable outcomes, which are part of the emergent and complex adaptation processes (Morchain, 2018). The values of justice, care and solidarity together could shape mutual encouragement, support and motivation to co-create fair and just adaptation strategies. Solving conflicting interests and viewing the present and the future, and simultaneously taking account of trade-offs between multiple courses of action require strengthening of deliberation and democratic tools based on these three value domains (Palmer, 2003; St. Clair, 2014). Guided by these three ethical values, debates and dialogues should happen on various possibilities and should engage with intersectional and intergenerational actors. The outcomes should be that each community actor participates proactively in forging caring solidarity for the well-being and sustainability of both human and non-human actors across diverse social-ecological systems. Having discussed the values guiding ethical adaptation to climate change, the next chapter details the various phases in the adaptive innovation process.

References

Adger, W.N. (2013). Emerging dimensions of fair process for adaptation decision-making, in J. Palutikof, S.L. Boulter, A.J. Ash, M.S. Smith, M. Parry, M. Waschka, and D. Guitart (Eds.). *Climate adaptation futures*, Chichester: Wiley-Blackwell, pp. 69–74.

Adger, W.N., Butler, C., and Walker-Springett, K. (2017). Moral reasoning in adaptation to climate change, *Environmental Politics*, 26(3), pp. 371–390.

Adger, W.N., Dessai, S., Goulden, M., Hulme, M., Lorenzoni, I., Nelson, D.R., Naess, L.O., Wolf, J., and Wreford, A. (2009). Are there social limits to adaptation to climate change? *Climatic Change*, 93(3–4), pp. 335–354.

Adger, W.N., Paavola, J., and Huq, S. (2006). Toward justice in adaptation to climate change, in A.W. Neil, J. Paavola, S. Huq, and M.J. Mace (Eds.). *Fairness in adaptation to climate change*, Cambridge, MA: The MIT Press, pp. 1–19.

Agrawal, A. (1995). Dismantling the divide between indigenous and scientific knowledge, *Development and Change*, 26(3), pp. 413–439.

Alexander, J.C. (2006). *The civil sphere*, New Delhi: Oxford University Press.

Arts, W., Muffels, R., and ter Mullen, R. (2001). Epilogue, in R. ter Mullen, W. Arts, and R. Muffels (Eds.). *Solidarity in health and social care in Europe*, London: Springer-Science + Business Media, B.V., pp. 463–477.

Averbuch, T. (2015). Entering, readiness, and contracting for dialogic organisation development, in G.R. Bushe and R.J. Marshak (Eds.). *Dialogic organisation development: the theory and practice of transformational change*, Oakland, CA: Berrett-Koehler Publishers, pp. 219–244.

Barnett, C. (2011). Geography and ethics: justice unbound, *Progress in Human Geography*, 35(2), pp. 246–255.

Barrientos, A. (2010). Protecting capability, eradicating extreme poverty: Chile Solidario and the future of social protection, *Journal of Human Development and Capabilities*, 11(4), pp. 579–597.

Bayertz, K. (1999). Four uses of 'solidarity', in K. Bayertz (Ed.). *Solidarity*, Dordrecht: Kluwer Academic Publishers, pp. 3–28.

Beck, U. (1992). *Risk society: towards a new modernity*, London: Sage.

Blau, J. (2017). *The Paris agreement: climate change, solidarity and human rights*, Cham: Palgrave Macmillan.

Bulkeley, H., Carmin, J., Broto, V., Edwards, G., and Fuller, S. (2013). Climate justice and global cities: mapping the emerging discourses, *Global Environmental Change*, 23, pp. 914–925.

Bulkeley, H., Edwards, G., and Fuller, S. (2014). Contesting climate justice in the city: examining politics and practice in urban climate change experiments, *Global Environmental Change*, 25, pp. 31–40.

Carmalt, J.C. (2011). Human rights, care ethics and situated universal norms, *Antipode*, 43(2), pp. 296–325.

Clark, H. (2009). Afterword: the chasing of nonviolence, in H. Clark (Ed.). *People power: unarmed resistance and global solidarity*, New York: Pluto Press, pp. 214–218.

COMEST. (2013). *Background for a framework of ethical principles and responsibilities for climate change adaptation*, Paris: UNESCO, 2013. Retrieved from http://unesdoc. unesco.org/images/0022/002264/226470E.pdf [Last accessed on 12 September 2019].

Conradson, D. (2011). Care and caring, in V.J. del Casino, Jr, M.E. Thomas, P. Cloke, and R. Panelli (Eds.). *A companion to social geography*, Oxford: Blackwell, pp. 454–471.

Cruz-Saco, M.A., and Zelenev, S. (2010). Introduction, in M.A. Cruz-Saco and S. Zelenev (Eds.). *Intergenerational solidarity: strengthening economic and social ties*, New York: Palgrave Macmillan, pp. 1–6.

de Beer, P., and Koster, F. (2009). *Sticking together or falling apart? Solidarity in an era of individualisation and globalisation*, Amsterdam: Amsterdam University Press.

Devereux, S., and Sabates-Wheeler, R. (2004). *Transformative social protection*. IDS Working Paper 232, Brighton: IDS. Retrieved from www.unicef.org/socialpolicy/files/ Transformative_Social_Protection.pdf [Last accessed on 3 November 2019].

Doherty, B., and Doyle, T. (2013). *Environmentalism, resistance and solidarity: the politics of Friends of the Earth International*, Basingstoke: Palgrave Macmillan.

Ferdinand, T. (2019). *Kenyan herders are switching from cattle to camels to adapt to climate change*, World Resources Institute, 9 July 2019. Retrieved from www.wri. org/blog/2019/07/kenyan-herders-are-switching-cattle-camels-adapt-climate-change [Last accessed on 17 July 2019].

Fincher, R., and Iveson, K. (2008). *Planning and diversity in the city*, New York: Palgrave.

Freire, P. (1970). *Pedagogy of the oppressed*, New York: Continuum.

Gardiner, S.M., and Weisbach, D.A. (2016). *Debating climate ethics*, New York: Oxford University Press.

Giddens, A. (1992). *The consequences of modernity*, Cambridge: The Polity Press.

Giddens, A. (1994). Living in a post-traditional society, in U. Beck, A. Giddens, and S. Lash (Eds.). *Reflexive modernisation: politics, tradition and aesthetics in the modern social order*, Cambridge: The Polity Press, pp. 56–109.

Glass, R.D., and Newman, A. (2015). Ethical and epistemically dilemmas in knowledge production: addressing their intersection in collaborative, community-based research, *Theory and Research in Education*, 13(1), pp. 23–37.

Gould, C.C. (2008). Recognition in redistribution: care and diversity in global justice, *Southern Journal of Philosophy*, 46(s1), pp. 91–103.

Grasso, M., and Feola, G. (2012). Mediterranean agriculture under climate change: adaptive capacity, adaptation, and ethics, *Regional Environmental Change*, 12, pp. 607–618.

Grieves, G., and Clark, M. (2015). Decolonising solidarity, *Overland*, 25 September. Retrieved from https://overland.org.au/2015/09/decolonising-solidarity/ [Last accessed on 22 August 2019].

Held, V. (2006). *The ethics of care: personal, political and global*, New Delhi: Oxford University Press.

Hetherington, T., and Boddy, J. (2013). Ecosocial work with marginalised populations: time for action on climate change, in M. Gray, J. Coates, and T. Hetherington (Eds.). *Environmental social work*, London: Routledge, pp. 46–61.

Hirota, M.Y. (2009). Complementary currencies as a method to promote the solidarity economy, in E. Kawano, T.N. Masterson, and J. Teller-Elsberg (Eds.). *Solidarity economy I: building alternatives for people and planet*, Papers and reports from the 2009 U.S. Forum on the solidarity economy, Amherst, MA: Centre for Popular Economics, pp. 75–86.

Hobson, K. (2006). Enacting environmental justice in Singapore: performative justice and the Green Volunteer Network, *Geoforum*, 37(5), pp. 671–681.

Hoggett, P., Mayo, M., and Miller, C. (2009). *The dilemmas of development work: ethical challenges in regeneration*, Bristol: The Policy Press.

Honneth, A. (1995). *The struggle for recognition: the moral grammar of social conflicts*, Cambridge, MA: The MIT Press.

ILO. (2001). *Social security: a new consensus*, Geneva: International Labour Office.

Indigenous Peoples of Mother Earth. (2012). Kari-Oca 2 declaration: indigenous peoples global conference on Rio+20 and Mother Earth, Kari-Oka Village, at Sacred Kari-Oka Puku, Rio de Janeiro, Brazil, 17 June 2012. Retrieved from http://www.inearth.org/docs/DECLARATION-of-KARI-OCA-2-Eng.pdf [Last accessed on 12 September 2019].

Jobin-Leeds, G. and AgitArte. (2016). Epilogue: solidarity – a gathering, in G. Jobin-Leeds and AgitArte (Ed.). *When we fight, we win: twenty-first-century social movements and the activists that are transforming our world*, e-book, New York: The New Press, pp. 421–440. ISBN 978-1-62097-140-6.

Joseph, D.D. (2017). Social work models for climate adaptation: the case of small islands in the Caribbean, *Regional Environmental Change*, 17, pp. 1118–1126.

Kivisto, P., and Sciortino, G. (2015). Introduction: thinking through the civil sphere, in P. Kivisto and G. Sciortino (Eds.). *Solidarity, justice and incorporation: thinking through the civil sphere*, New York: Oxford University Press, pp. 1–31.

Klepp, S., and Chavez-Rodriguez, L. (2018). Governing climate change: the power of adaptation discourses, policies, and practices, in S. Klepp and L. Chavez-Rodriguez (Eds.). *A critical approach to climate change adaptation: discourses, policies and practices*, London: Routledge-Earthscan, pp. 3–34.

Kolers, A. (2016). *A moral theory of solidarity*, Oxford: Oxford University Press.

Krol, M. (2016). Introductory remarks, in J. Kottan (Ed.). *Solidarity and the crisis of trust*, Gdansk: European Solidarity Centre, pp. 7–12.

Kronlid, D.O. (2014). *Climate change adaptation and human capabilities: justice and ethics in research and policies*, New York: Palgrave Macmillan.

Kwan, C., and Walsh, C.A. (2015). Climate change adaptation in low-resource countries: insights gained from an eco-social work and feminist gerontological lens, *International Social Work*, 58(3), pp. 385–400.

Leichenko, R., and Silva, J. (2014). Climate change and poverty: vulnerability, impacts and alleviation strategies, *WIREs Climate Change*, 5, pp. 539–556.

Light, A., and Rolston, H. (Eds.). (2003). *Environmental ethics: an anthology*, Oxford: Blackwell.

Masschelein, J. (2010). Educating the gaze: the idea of poor pedagogy, *Ethics and Education*, 5(1), pp. 45–53.

Mearns, R., and Norton, A. (2010). Equity and vulnerability in a warming world: introduction and overview, in R. Mearns and A. Norton (Eds.). *Social dimensions of climate change: equity and vulnerability in a warming world*, Washington, DC: World Bank, pp. 1–44.

Moellendorf, D. (2012). Climate change and global justice, *Wiley Interdisciplinary Reviews: Climate Change*, 3(2), pp. 131–143.

Mohanty, C.T. (2003). *Feminism without borders: decolonising theory, practicing solidarity*, Durham, NC and London: Duke University Press.

Morchain, D. (2018). Rethinking the framing of climate change adaptation: knowledge, power, and politics, in S. Klepp and L. Chavez-Rodriguez (Eds.). *A critical approach to climate change adaptation: discourses, policies and practices*, London: Routledge-Earthscan, pp. 55–73.

Naidoo, L. (2017). Women's boat to Gaza: an international solidarity mission, in A. von Kotze and S. Walters (Eds.). *Forging solidarity: popular education at work*, Rotterdam: Sense Publishers, pp. 49–58.

Olthuis, J.H. (2000). Exclusion and inclusions: dilemmas of difference, in J.H. Olthuis (Ed.). *Towards an ethics of community: negotiations of difference in a pluralist society*, Ontario: Wilfred Laurier University Press, pp. 1–10.

Oosterlynck, S., Schuermans, N., and Loopmans, M. (2017). Beyond social capital: place, diversity and solidarity, in S. Oosterlynck, N. Schuermans, and M. Loopmans (Eds.). *Place, diversity and solidarity*, London: Routledge, pp. 1–18.

Page, E.A. (2007). *Climate change, justice and future generations*, Cheltenham, UK and Northampton, MA: Edward Elgar Publishing.

Palmer, C. (2003). An overview of environmental ethics, in A. Light and H. Rolston (Eds.). *Environmental ethics: an anthology*, Oxford: Blackwell, pp. 15–37.

Parrott, L. (2010). *Values and ethics in social work practice*, Exeter: Learning Matters Ltd.

Peeters, J. (2012). A comment on 'climate change: social workers' roles and contributions to policy debates and interventions', *International Journal of Social Welfare*, 21, pp. 105–107.

Pelling, M. (2011). *Adaptation to climate change: from resilience to transformation*, London: Routledge.

Ranganathan, M. (2017). The environment as freedom: a decolonial reimagining. *Social Science Research Council Items*, 13 June. Retrieved from https://items.ssrc.org/just-environments/the-environment-as-freedom-a-decolonial-reimagining/ [Last accessed on 13 August 2019].

Ranganathan, M., and Bratman, E. (2019). From urban resilience to abolitionist climate justice in Washington, DC, *Antipode*, pp. 1–23. ISSN 0066-4812. Retrieved from https://onlinelibrary.wiley.com/doi/epdf/10.1111/anti.12555 [Last accessed on 8 January 2020].

Ravi, R., and Vanak, A.T. (2019). Why has drought hit the Maldharis of Kutch so hard this year? *The Hindu*, 11 May 2019. Retrieved from www.thehindu.com/sci-tech/energy-and-environment/drought-in-a-desert-why-has-drought-hit-the-maldharis-so-hard-this-year/article27090863.ece [Last accessed on 2 November 2019].

Rawls, J. (1971). *A theory of justice*, Cambridge, MA: The Belknap Press of Harvard University.

Reitan, R., and Gibson, S. (2012). Climate change or social change? Environmental and leftist praxis and participatory action research, *Globalizations*, 9(3), pp. 395–410.

Reyes, J.A.C. (2016). *Disaster citizenship: survivors, solidarity, and power in the progressive era*, Chicago, IL: University of Illinois Press.

Robinson, M. (2019). *Climate justice: a man-made problem with a feminist solution*, London: Bloomsbury.

Ruddick, S. (1995). Injustice in families: assault and domination, in V. Held (Ed.). *Justice and care: essential readings in feminist ethics*, Boulder, CO: West-view Press pp. 203–223.

Santha, S.D., Jaswal, S., Sasidevan, D., Khan, A., Datta, K., and Kuruvilla, A. (2016). Climate variability, livelihoods and social inequities: the vulnerability of migrant workers in Indian cities, *International Area Studies Review*, 19(1), pp. 76–89.

Schlosberg, D., and Collins, L. (2014). From environmental to climate justice: climate change and the discourse of environmental justice, *Wiley Interdisciplinary Reviews: Climate Change*, 5(3), pp. 359–374.

Scholz, S.J. (2008). *Political solidarity*, University Park, PA: The Pennsylvania State University Press.

Sen, A. (2009). *The idea of justice*, London: Allen Lane, Penguin books.

Sevenhuijsen, S. (1998). *Citizenship and the ethics of care: feminist considerations on justice, morality and politics*, London: Routledge.

Shiva, V. (2000). *Stolen harvest: the hijacking of the global food supply*, New Delhi: India Research Press.

Smith, S.J. (2009). Everyday morality: where radical geography meets normative theory, *Antipode*, 41(1), pp. 206–209.

St. Clair. (2014). The four tasks of development ethics at times of a changing climate, *Journal of Global Ethics*, 10(3), pp. 283–291.

ter Mullen, R., Arts, W., and Muffels, R. (2001). Introduction to the volume, in R. ter Mullen, W. Arts, and R. Muffels (Eds.). *Solidarity in health and social care in Europe*, Dordrecht, Boston, MA and London: Kluwer Academic Publishers, pp. 1–11.

Till, K.E. (2012). Wounded cities: memory-work and a place-based ethics of care, *Political Geography*, 31, pp. 3–14.

Tronto, J.C. (1993). *Moral boundaries: a political argument for an ethic of care*, London: Routledge.

Uncapher, D., and Yvellez, C. (2019). Climate adaptation needs to put human rights above property values, *Truthout*, 14 July 2019. Retrieved from https://truthout.org/articles/climate-adaptation-needs-to-put-human-rights-above-property-values/ [Last accessed on 16 July 2019].

Vielmas, S. (2017). Building a movement for the right to education in Chile, in A. von Kotze and S. Walters (Eds.). *Forging solidarity: popular education at work*, Rotterdam: Sense Publishers, pp. 193–202.

von Kotze, A., and Walters, S. (2017). Introduction, in A. von Kotze and S. Walters (Eds.). *Forging solidarity: popular education at work*, Rotterdam: Sense Publishers, pp. 1–14.

Walters, S., and Butterwick, S. (2017). Moves to decolonise solidarity through feminist popular education, in A. von Kotze and S. Walters (Eds.). *Forging solidarity: popular education at work*, Rotterdam: Sense Publishers, pp. 27–38.

Whyte, K.P. (2019). Way beyond the lifeboat: an indigenous allegory of climate justice, in K.-K. Bhavnani, J. Foran, P.A. Kurian, and D. Munshi (Eds.). *Climate futures: reimagining global climate justice*, London: Zed Books, pp. 10–18.

Whyte, K.P., and Cuomo, C. (2017). Ethics of caring in environmental ethics: indigenous and feminist philosophies, in S.M. Gardiner and A. Thompson (Eds.). *The Oxford handbook of environmental ethics*, New York: Oxford University Press, pp. 234–247.

Williams, M. (2016). Searching for actually existing justice in the city, *Urban Studies*, pp. 1–15. DOI: 10.1177/0042098016647336.

Williams, M.J. (2017). Care-full justice in the city, *Antipode*, 49(3), pp. 821–839.

Young, I.M. (1990). *Justice and the politics of difference*, Princeton, NJ: Princeton University Press.

Young, I.M. (2004). Responsibility and global labor justice, *Journal of Political Philosophy*, 12, pp. 365–388.

5 Adaptive innovation phases

Introduction

The aim of this chapter is to describe the major phases in the adaptive innovation model. A regular approach for innovation with respect to simple systems has been a three-step cycle of planning, implementation and evaluation. However, to deal with complex systems and problems such as climate change adaptation, the adaptive innovation model is envisaged as a suitable alternative. Adaptive innovation is an iterative and reflective process among diverse actors who are involved in a shared and interactive frame of analysis, ideation and practice. The key assumption of this model is that community actors involved in climate change adaptation can enhance their adaptive capacities through collective, dialogic and reflective processes of innovation. The ethical values embedded in the adaptive innovation model also provide scope to include the marginalised and subjugated actors in adaptation planning and decision-making processes. The whole process followed in adaptive innovation also foresees in keeping alive the aspirations, hope, trust, networks, capacities and willingness of actors participating in the change-making process. An introductory understanding to the six phases of adaptive innovation was provided in Chapter 3. The six phases of adaptive innovation are situational analysis, micro-mobilisation, dialogic ideation, action framing, piloting and emergence (refer to Figure 5.1).

Each of the six important phases in adaptive innovation model is described in the following sections.

Phase 1: Situational analysis

Situational analysis is a process of elaborately understanding the changes happening to a particular social-ecological system and its impact on humans and non-humans. It involves gathering a preliminary understanding on the vulnerability contexts, livelihood practices, adaptation trends and other key issues affecting diverse community actors in a given social-ecological system. There are three important steps that could guide a situational analysis, namely (a) scoping review, (b) participatory mapping and (c) analysis of drivers and barriers to collectivisation, innovation and adaptation. Though these steps are

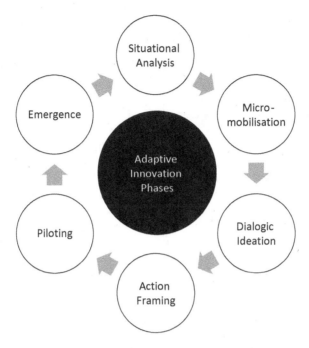

Figure 5.1 Adaptive innovation phases.

presented in a linear way for the sake of better understanding to the reader, one has to anticipate that in practice these processes are non-linear, intertwined and dynamic in nature. Each of these steps is explained in detail as follows.

Scoping reviews

Scoping reviews enable us to gather a broader and initial understanding of the situation. It involves exploring the respective situation at hand through preliminary field visits and simultaneous review of literature. The scoping review has to be flexible enough that we are able to explore diverse domains pertaining to the dynamics of the social-ecological system. The aim is to gather a quick overview of what has happened before, and what is happening now in the context of changes in the social-ecological system, including climate change, livelihood practices and related institutional contexts. It is also important to explore the narratives that each actor shares about these changes and their impacts. Analysis of trends, key local information and historical changes are important components of the scoping review. It is also imperative to look out for existing gaps in our capacities to understand the present situation. Specifically, we have to gather some basic understanding of the social-ecological and livelihood conditions that shape the vulnerability contexts of diverse community actors.

A scoping review acts as a framework for synthesising available published and grey literature on the subject matter associated with the practice situation in focus (Arksey and O'Malley, 2005; Gough et al., 2012; Peterson et al., 2017). These include theoretical and narrative reviews, published and unpublished empirical research papers and reports, crucial statistical information such as socio-demographic data and grey literature published by government agencies, non-governmental organisations and other similar sources (Peterson et al., 2017). Browsing across all national and regional newspapers and magazines including new social media to unearth their reporting on local level environmental and livelihood transitions including incidences of extreme hazard events (that are indicative of climate change) and their impacts on the particular social-ecological system will add value to the scoping review. We can also gather expert opinion within and outside the community to complement the scoping review. During our preliminary field visits, we need to explore and probe the shocks, uncertainties and changes happening to the day-to-day life and livelihood practices of community actors. These preliminary field visits are also crucial for us to build rapport and trust with key community actors. Our ability to interact with a wide range of community actors and quickly gather a deeper understanding of the history, cultural and behavioural characteristics of actors therefore becomes crucial. Nevertheless, we should be ready to refine these understandings based on the emergent characteristics of each situation. Insights can be gathered through oral histories and in-depth interviews.

The scoping review has to pay specific attention to the analysis of vulnerabilities and capacities of both human and non-human actors in the social-ecological system. Factors such as average landholding size, contribution of livestock and allied farming systems to overall share of livelihoods, changes in cropping system, design and scale of technological innovations, institutional change, transitions in resource use and access, migration trends and the nature of social capital could indicate the nature of adaptive capacities of community actors (Ojha et al., 2014). Specific attention has to be given to the local knowledge systems, perspectives and practices of subjugated actors. Their situated knowledge corresponding to respective social positions, their access to resources and ability to diversify their livelihoods have to be adequately explored, as they determine the adaptive capacities of these actors. The findings of such a review can be shared or presented to the community actors and other stakeholders for feedbacks, comments and towards generating adequate momentum for dialogue and action. These presentations have to be carried out in simple local language, so that all concerned actors are able to clearly comprehend the situation (Peters et al., 2015; Peterson et al., 2017). The dialogue thus generated should be carried forward through follow-up sessions, including interactions during the participatory mapping and analysis of drivers and barriers.

Participatory mapping

Participatory mapping is a practice-based interactive approach that aims at spatially representing the understanding, knowledge and experiences of local community actors. It can represent perceptions, experiences and understanding of diverse community actors that are located within certain temporal, spatial and social contexts (Chambers, 1994, 1997). The primary aim of participatory mapping in this phase of adaptive innovation is to map the transitions in the social-ecological system. The aim is also to identify and describe our primary impact groups, the changes in their social and ecological worlds, their livelihood practices and other relevant stakeholders. We could develop some clarity on the diversity and dynamics of resource users and corresponding resource use changes over a period of time.

While doing the mapping exercise, specific attention has to be given to the analysis of livelihood contexts and coping strategies of subjugated actors and other vulnerable groups, who are deprived politically and economically, and whose voices are often not recognised and remain unheard. Indicators of adaptive capacity such as livelihood assets, local institutional arrangements and ongoing adaptation strategies have to be explored during the mapping process (Jones et al., 2010; Piccolella, 2013). During the mapping process, our interactions can also explore aspects such as major shocks and uncertainties, demographic transitions, local economic trends and institutional changes, individual and collective responses to environmental and economic risks, transitions in asset use and its implications on ecosystem services. Studies have demonstrated the utility of these maps in facilitating the planning of community-based adaptation strategies to climate change and extreme hazard events (Barkved et al., 2014; Di Gessa et al., 2008; Piccolella, 2013).

The process of participatory mapping would enable us to nurture collective action and dialogue among community actors. These exercises would also help in triggering reflections, discussions and problem framing with regard to the value and ongoing dynamics of social-ecological systems (Beichler, 2015). They also facilitate collective dialogue and learning about extreme climatic events and its impact on vulnerable groups (Barkved et al., 2014). Local variations in perceptions and experiences of diverse community actors in specific social positions on climate variability and their differential exposure to extreme hazard events can also be captured using participatory mapping methods with specific focus groups. It is presumed that during this phase, we would have developed meaningful rapport and some level of trust with community actors and other relevant stakeholders. Without adequate rapport and trust being developed with the concerned actors, situational analysis and the follow-up process have limited meaning. Trust-based relationship building is thus a prerequisite for adaptive innovation. Participatory mapping has the potential to develop innovative pathways to problem solving such as the creation of inventive institutional processes or the transfer of new technologies (Piccolella, 2013). With advancements in remote sensing and Geographic Information Systems (GIS), participatory mapping can also be combined with both the tools of modern cartography and traditional participatory methods to represent the spatial knowledge of local communities (Barkved et al., 2014; Guru and Santha, 2013).

Participatory analysis of drivers and barriers

As part of the situational analysis, we also have to explore the explicit and tacit factors that could enable and impede collective action and innovation in climate change adaptation. The participatory mapping process therefore needs to be followed by the analysis of drivers (or enablers) and barriers (or constraints) to climate change adaptation. A driver to adaptation is an enabling factor, while a barrier acts as an impediment to specific forms of adaptation for a particular group of actors (Eisenack et al., 2014; Rendón and Gebhardt, 2016). However, a driver or barrier can be valued differently by different actors (ibid).

We could encounter diverse enabling and constraining factors associated with effectively motivating and organising community actors as a collective. Contextual factors such as access to resources, decision-making arrangements and prevailing economic and political ideologies could either enable or constrain adaptation efforts; varying from one set of actors to another based on their situatedness and social positions. Components of culture, legal and governance systems, health of ecosystems and their services, migration processes and social networks are also critical factors to consider, as they will influence how drivers and barriers emerge in local places (Bekkers et al., 2013; Shackleton et al., 2015; Wiederkehr et al., 2018). For example, a strong history of environment conservation by local communities and its legitimation by the government could become a driver for taking adaptation decisions (Alemagi et al., 2006). Self-determination of community actors to develop their own adaptation strategies are found to be quite successful in developing appropriate solutions (Vanni, 2014). Our role in such contexts is to facilitate interaction and shared learning rather than directly persuading participant actors on the adaptation strategies to be adopted (ibid).

Risk perception is yet another important driver to adaptation (Dilling et al., 2017). Other aspects such as simplicity of adaptation measures, possibility of income generation, non-violence, technical feasibility, financial viability, social equity and cultural acceptability could also facilitate adaptation (Mondal et al., 2010). In this regard, long-term sustainability of innovations in adaptation is often determined by resources provided by the state and can be constrained or enabled by the prevailing regulatory environments (MacCallum, 2013). Constraints related to knowledge, awareness and technology; the physical environment; biological tolerances; economic and financial factors; human resources; social and cultural factors; and governance and institutional processes could act as barriers to ethical adaptation (Klein et al., 2014; Mondal et al., 2010). Lack of collaborative partnerships, absence of political will and leadership, lack of public awareness, inadequate monetary and human resources and poor implementation of policies can act as barriers to adaptation (Ireland and Thomalla, 2011; Oulahen et al., 2018; Rendón and Gebhardt, 2016; Thaler et al., 2019). Factors such as heterogeneity of community actors, lack of trust, free riding, dynamic resource units, ambiguous resource boundaries and lack of legitimation of local institutions could also constrain the sustainable management of social-ecological systems (Ostrom, 1990).

Barriers can also emerge from self-interest-driven behaviour of community actors (Gawel et al., 2012). Higher transaction costs may also act as important barriers to collective action in the initial stages of adaptation (Ostrom, 1990). The discontinuities resulting out of knowledge interfaces between community actors and outside experts could also end up as barriers to innovation and adaptation (Long and Long, 1992; Rendón and Gebhardt, 2016). Path dependency could also emerge as a deep barrier to adaptation (Barnett et al., 2015). Adhering to past beliefs, structures and traditions, path dependency manifests as resistance to changing the way things have always been done (ibid). Resistance exists even if the outcomes of the response were maladaptive (ibid). The issues of population vulnerability and social exclusion will therefore remain as critical factors, which also would influence who experiences what barriers (Barnett et al., 2015; Shackleton et al., 2015). Oppressive cultural norms across different institutional layers could constrain ethical adaptation at the local level (Jones, 2010). Therefore, it is also evident that meaningful, well-intended structural and societal transformations are necessary for adapting to climate change. Hence, this also requires starting enquiring about the root causes of these barriers, who they affect most and what is needed to overcome them (Shackleton et al., 2015). An understanding on the nature of the barrier, its source(s) and the location of influence could provide future pathways to overcome the constraints (Ekstrom et al., 2011).

Apart from the spatial and segmental dimensions, the analysis could also look at the barriers and drivers with a sector focus. These would include water, agriculture, fisheries, health, tourism, human settlements and built environment, coastal zones, energy, finance and insurance. It will be crucial to examine and deliberate on the impacts of climate variability and extreme hazard events on the sectors in which the livelihoods of specific community actors are dependent on. For example, we could facilitate discussions on phenomena such as water scarcity and its implications on gender roles. While deliberating on agriculture, participant actors should be able to reflect on the day-to-day life situations of vulnerable farm families and the impact of climate change on their livelihood diversification strategies, health of livestock, nature of pest attack on crops, exclusion and inclusion in the agro-value chain, access to farm inputs and the strength of present coping strategies. Such enquiries can be continued with other allied sectors as well. Other follow up questions depending on the contexts could be related to the nature of water demand, flood-drought frequencies, coastal flooding and saline intrusion, and prevailing strategies to access infrastructure services, crop management, livestock management, soil and water management, income diversification, food provision, maintaining social networks, migration or barriers to access humanitarian aid, as the case may be (Wiederkehr et al., 2018; Wreford et al., 2017). Apart from these, it is also important to discuss and analyse the impact of climate change on the health of people, with specific reference to infectious diseases, food security and malnutrition, water and sanitation and so on.

Exploring the primary drivers for innovation also gains significance in the context of adaptation planning. The specific challenges will be towards strengthening drivers, and to overcome barriers. Innovations in climate change adaptation could face different drivers, and barriers at different stages. The role of government and market in creating incentives for innovation could be explored. More importantly, the participant actors themselves could reflect on how they are driving innovation in each sector (Foxon et al., 2005). In addition, the drivers and barriers to adaptation are dynamic and context specific, therefore, their impacts change in the course of the adaptation process (Rendón and Gebhardt, 2016). The perception of what factor is a driver, or a barrier depends on the specific context and on the perspective of the actor, who judges its impact (ibid).We have to facilitate participant actors to explore whether the proposed adaptation scenario is linked to more fundamental social transformations in the long term or whether it stipulates certain immediate actions for risks at hand. Innovation projects tend to succeed when local demand is made explicit and local community actors are involved in the processes from the very beginning (Mondal et al., 2010). Inducing a sense of ownership is in itself a driver for innovation and sustainable adaptation.

It is envisaged that by the end of this phase, we would have identified specific groups of community actors who have a stake and are interested to be part of the innovation process in designing and implementing just adaptation strategies. It is advisable at this stage to carry out an in-depth stakeholder analysis, identifying crucial actors from both inside and outside the community, critically reflecting on their social positions, interests and aspirations and introspecting on possible implications of their participation (or non-participation) in the adaptation projects. Though the involvement of diverse actors with different stakes would remain as a challenge, we may have to still strive towards nurturing partnerships amongst relevant actors across the public, private and third sectors so as to stimulate innovation (Martinelli, 2012). Hence, trust creation between these actors is an important responsibility of ours (Jalonen and Juntunen, 2011). Leadership, strategic thinking, resourcefulness, creativity, collaboration and effective communication are all other crucial factors to design appropriate adaptation strategies (Ekstrom et al., 2011).

S.F.F.5.1 Narratives on situational analysis.

Situational Analysis is perhaps one of the most crucial but challenging phases in the adaptive innovation project. Our skills as social workers to remain empathetic to the people with whom we work with and at the same time understand their situation in-depth to bring about meaningful changes is crucial in this regard. A few years ago, I was vested with the responsibility of carrying out a situational analysis with a fishing community that was very vulnerable to coastal erosion. The village was located between a fishing harbour and an estuary. There were around 300 households in the village and most of them used to depend on the artisanal fisheries sector for their survival. The week before I visited this village for the first time, tidal waves and sea surges had destroyed three houses in this village. More than 100 families were relocated to the nearby school. For the first few days, I had to silently observe the situation with empathy. I did have discussions with some of the key informants who were able provide more details on the present event as well as on the historicity of the village. As things began to settle down and families started returning to their houses, I had in-depth interactions with few willing participants. Door-to-door household visits and conversational interviews were very helpful during this phase. Gradually, we came together to collectively deliberate and reflect on the situation. These processes were very organic in nature and nothing was enforced from outside. Transect walks and informal group discussions did help us to gather more understanding on the causes of hazards and needs and aspirations of people with respect to resource management. We had discussions separately with men, women, children, youth and the elderly. The officials of the local governing body, church and the Fisheries Department were also present during some of these meetings. In the meantime, I also explored other grey literature and official documents pertaining to the historicity and development plans of the village and nearby coastal resources.

The situational analysis revealed that the fishing village was vulnerable to different coastal hazards due to man-made factors. More than 100 households were susceptible to flooding due to the impact of tidal waves and surges. The whole village is vulnerable to coastal erosion and landward intrusion of saline ground water. Drinking water was becoming scarce due to coastal inundation and the subsequent salinisation of drinking water sources. The government had constructed sea walls out of huge boulders to check coastal erosion. However, this was not a sustainable mechanism and it further affected the livelihood security of fisherfolk. An elderly fisherman narrated his ordeal as follows:

> These days, the climate is changing drastically both in the land and at the sea. The wind and the waves have grown stronger. The storm surges have become more frequent and intense. As the directions of the wind have become unpredictable and overpowering, our catamarans often end up in the wrong shore. We are also unable to cast our nets properly. The waves are so huge that some of our catamarans have endured irreparable damage. Nevertheless, as we know the sea, we can survive all these threats. But what threatens us the most today is coastal erosion! All our lands are gone! Many have become homeless! Whatever assets we have acquired after struggling in the sea, the sea is claiming them back. Earlier, the "southern sea" (South-West Monsoon) used to cause coastal erosion. That was a natural phenomenon and predictable. Whatever land was lost during that time, we used to get it back with the arrival of the "northern sea" (North-East Monsoon). Today, we cannot predict when the erosion happens. It occurs everyday, every moment, even when the sea is calm!

The Integrated Rural Health and Development Project (IRHDP) is a field action project of TISS. The project aims at improving the wellbeing of tribal communities in Aghai village of Maharashtra, India. I began my association with the project in the year 2014, when the Project Director requested support to strengthen climate-resilient livelihoods in the project villages. The project facilitated an action research approach. In the beginning phase of situational analysis, we carried out a socio-historical analysis of the village through key informant interviews, oral histories and in-depth group discussions. During this phase, community participation was quite limited, and the processes were largely facilitated by the project team. However, we could observe that these tribal households were exploited by intermediaries in their seasonal agricultural transactions; and they were excluded by the government departments in providing accurate information and effective services as mandated by the State. Gradually, participatory interactions with community members were facilitated. We were engaged in a series of household visits, community profiling, transect walks, resource and risk mapping, participatory livelihood analysis sessions and group meetings to understand the living conditions of community actors in these villages. This phase of situational analysis revealed that the vulnerability contexts of tribal communities in the project area had worsened with climate variability such as erratic monsoons, sporadic rainfall, drought and water scarcity. Other secondary impacts that we identified were decline in crop productivity, large-scale out-migration of men to cities in search of employment and specific health issues. Further, these forces have burdened the women with additional work commitments both at home and in the agricultural fields.

Phase 2: Micro-mobilisation

The next phase in adaptive innovation is mobilising diverse community actors towards creating a shared vision and acting upon it. Micro-mobilisation is a dynamic and ongoing process. It requires a sense of visioning, consensus building and legitimation by the community actors and other relevant stakeholders. The primary aim of adaptive innovation is to enhance the participation of diverse community actors in the adaptation processes (Ward, 2016). The willingness of community actors to participate alone would not lead to actual participation. The political and cultural contexts constrain or enable social mobilisations (Viterna, 2013). Our role is to enable community actors to overcome certain barriers that prevent them from participation and at the same time create a motivating environment for collective action (Klandermans, 1984).

Community actors in a collective get motivated when they are able to relate to and develop an ideological affinity and shared vision about the future. In this regard, we could begin our work with those community actors who share similar concerns towards climate change impacts and believe in the need for adaptation. Community actors possess multiple subjectivities and identities and therefore will relate more to those identities and social categories (e.g. class, caste, race, gender, language, occupation and so on) that they are more affiliated with in a particular social world, as compared to others. Micro-mobilisation, therefore involves a frame alignment of values and interests among diverse actors, and frame resonance in building knowledge consensus and mediating power structures (Snow and Benford, 1988). Creating frame alignment toward ethical adaptation involves the convergence of ideas, beliefs and commitment among diverse community actors. This necessitates that we engage primarily with the identity, needs and aspirations of community actors (Viterna, 2013). Identities are the roles that each actor occupies or associates to in specific social worlds. Each social encounter in adaptive innovation may reinforce or reshape the meanings of actors' identities and their associations with the respective social worlds.

Examining the implications of these processes and negotiating with community actors to actively participate in climate change adaptation is therefore a challenging, but inevitable task for us. Moreover, adapting to complex environmental problems such as climate change involves coordinating actions across diverse social-ecological domains. These domains could spread across diverse sectors such as agriculture, land use, water management, forestry or fisheries. This also implies that actions have to be coordinated on diverse resource systems such as farmlands, water bodies, forests, fisheries and several other common property resources such as grazing lands. Micro-mobilisation is thus challenging even at village level, as several community actors with diverse interests, livelihoods and priorities of different uses of land and water are required to work together (Tucker et al., 2013). Facilitating meaningful exchange of information, sharing ideas, planning and solving problems as a collective is a way forward to address these challenges (Pound and Conroy, 2017). Applying meaningful strategies such as developing shared visions embedded with the emotions, narratives and actions of community actors could gradually enable them to internalise their identity and align their frames with the adaptive innovation process (Viterna, 2013). Our understanding of why actors opt for particular actions and narratives in a particular social encounter, and its possible intended and unintended consequences therefore gain significance.

The micro-mobilisation processes also involve the building and strengthening of actor networks across different levels of intersectional backgrounds; sharing common goals, ideas and aspirations to deal with climate change and extreme hazard events. These network spaces could be gradually nurtured as innovation platforms that are shaped through dialogue and caring practices. These innovation platforms should have the potential to allow diverse community actors to come together to address issues of mutual concern and interest (Bailie et al., 2018).

They have to be envisaged as spaces to facilitate better communication, negotiation, coordination and knowledge sharing among diverse actors (Raj and Bhattacharjee, 2017; Swaans et al., 2014). If designed properly, the innovation platforms could facilitate quick and continuous feedback from concerned actors at all stages of adaptive innovation. Such an approach could facilitate the timely integration of new knowledge into the innovation process using feedbacks from experiential learning and reflective practice (Nyikahadzoi et al., 2012). These are essential pre-requisites to work with complex systems as well.

The innovation platforms proposed can take any institutional form. They can be informal, formal or hybrid collectives. These need not be some new form of institutions but can be built on the foundations of existing local-level institutions. Before deciding on a particular form of innovation platform as a space for adaptive innovation, we have to cross-check whether any other institutional forms could facilitate adaptive innovation in a better way. Resilient organisations that are already present at the local level could emerge as appropriate institutions mediating innovation and adaptation. These alternatives could be an informal collective, a self-help group, a micro-credit group, a producer cooperative, a trade union, farmer field schools, local self-governance bodies or any other people-centred institution. However, care has to be taken that these institutions function within a frame of fairness or justice, care and solidarity.

Improvements to complex social-ecological systems depend on broader institutional innovations that are pluralistic and context-specific. An overemphasis on the rigidity of membership and organisational structure may in itself become a barrier to collectivisation. Instead, our focus should be on nurturing and building trust among diverse actors and in continuously evolving and reflecting upon the mutually agreed upon roles and responsibilities. And at the same time, we should facilitate the membership building processes in such a way that the vulnerable and marginalised groups get due representation in the respective platform. The micro-mobilisation processes can be strengthened through a series of individual and group meetings, and participatory workshops with different community actors. The individual interviews and group discussions can be held with specific homogenous groups of community actors, while the participatory workshops can be organised with heterogeneous groups. The initial meetings will be generally oriented towards creating awareness and sensitising potential members about the need for collective action in climate change adaptation. Once the community actors are organised as a group or a collective entity, we should gradually proceed further with the adaptive innovation strategy of developing shared visions. Steps such as shared visioning and risk and vulnerability analysis could strengthen the micro-mobilisation process. These steps are discussed in detail in the following sections.

Shared visioning

Shared visioning is about developing value-based actionable framework that could nurture co-creation and collaborative engagement amongst diverse actors with different perspectives and experiences (Schusler et al., 2003). The visioning for climate change adaptation has to begin with the assumption that the environment is a meta-capability for both human and non-human wellbeing (Holland, 2012). The underlying thought that would guide the visioning process is 'how each actor as individuals and as collectives would like to visualize their future – in the context of a complex and uncertain world?' The capacity to vision a safe and secure future also could determine the resilience of the social-ecological system (Beaulieu et al., 2015). Visioning involves the interactive and dialogical processes where diverse community actors align their values and interests to create a safe and secure future for them (Dolleris, 2011).

Shared visioning is a process by which community actors express their hopes and aspirations, and at the same time develop consensus about an ideal future (Dolleris, 2011). Visioning should be based on the situated knowledge of community actors rather than thrusting a single expert-driven knowledge system (Nyikahadzoi et al., 2012). It should be primarily based on the needs of community actors. Further, any community-based visioning process has to be embedded in the context (Swaans et al., 2014). The aim should be to stimulate incremental change through diverse bundles of innovation that are rooted in reflective practice (ibid). Community actors involved in visioning should maintain an organic connect to the context being outlined and the issues being discussed. Participation of diverse actors and their varied understanding of the context has to be recognised and deliberated upon. Each actor should have the willingness to listen and be empathetic to others' views and their situated contexts. Some of the important characteristics that foster effective visioning processes are open communication, diverse participation, unrestrained thinking, constructive conflict, democratic structure, multiple sources of knowledge, extended engagement and facilitation (Schusler et al., 2003).

The visioning process is also about creating an atmosphere of generating a shared commitment. We, as social workers, could facilitate the process of securing mutual commitment and developing a sense of ownership from all concerned actors. Specific focus has to be given to voices, hopes and aspirations of marginalised and subjugated actors. Other actors situated higher in the power hierarchy should consider it as their obligation to address these aspirations (Chambers, 1983). We also need to be alert to the intersectional variations in these voices; and reflective to the power dynamics embedded in the visioning processes. For, shared visioning may not always result in the direct empowerment of subjugated actors (Nieto-Romeroa et al., 2016). In contrast, the dominance by a select few actors in the visioning processes could exclude the subjugated actors as well (Krzywoszynska et al., 2016). Visioning therefore has to happen as a shared exercise where each actor becomes aware of their own reality and fellow actors' situatedness as well; and how each one of them understand these realities and expresses them in their day-to-day practice (Steyaert et al., 2007).

Adaptive innovations happen through the co-creation of shared meanings among participant actors. In the context of visioning, it involves a collective sense of reflecting on what is important, and why (Senge et al., 2010). These narrations and reflections have to be therefore weaved around the life stories and lived experiences of diverse actors, emphasising on why their voices have to be heard and how they fit into the larger world (ibid). In adaptive innovation, we are not aiming at a single-vision statement. Instead diverse standpoints could emerge, representing the varied lived experiences and aspirations of different community actors. The diversity of the visions situated within the embodiment and historicity of each actor has to be empathically recognised. If this is not practiced, then only certain visions stated by dominant actors would be promoted constraining the quality of relationships in the innovation process (Senge, 1990). Unless the diversity is harmonised, deeper and common visions will not emerge (ibid). The non-recognition of the variations in visions can also result in the framing of ineffective adaptation strategies (Adger et al., 2009). Our role is to facilitate meaningful deliberations among diverse actors to understand the underlying values shaping their preferences while visioning adaptation situations (Beaulieu et al., 2015).

Building shared visions require time, care and strategy (Senge et al., 2010). We should not facilitate the processes of shared visioning in a haste. Adequate time has to be allocated for the vision-building process. The ultimate ability of the vision to develop as a vibrant force of change happens only when participant actors truly believe that they can shape their future (Senge, 1990). And at the same time, it should be flexible enough to support incremental change (Swaans et al., 2014). We could triangulate using different participatory methods to capture the meanings attributed by participant actors on their envisioned future. These meanings will vary from one group of actors to another based on diverse contextual factors including the historical and social positions of participant actors. We have to continually listen for the emerging purpose, and consistently gauge the ups and downs in the interests and needs of the actors (Senge et al., 2010). Transparent mechanisms to share information and communication would provide confidence among actors to actively participate in the visioning process (Swaans et al., 2014). These mechanisms of openness and genuine caring also determine the quality and power of the vision (Senge et al., 2010).

Risk and vulnerability analysis

Actors participating in the visioning process should also be aware of the risk and vulnerabilities of community actors. Certain questions pertaining to adaptation have political implications linked to questions of power and resources, social inequality and access to decision-making and information that makes the task of adaptation and climate justice even more difficult (Klepp and Chavez-Rodriguez, 2018; Pelling, 2011). In this context, co-exploring the nature of vulnerability together with community actors would enable us to understand how these actors are differentially exposed to the impacts of climate change.

Vulnerability could be defined as

> the degree to which a system is susceptible to, and unable to cope with, adverse effects of climate change, including climate variability and extremes.

Vulnerability is a function of the character, magnitude, and rate of climate change and variation to which a system is exposed, its sensitivity, and its adaptive capacity.

(IPCC, 2007: 27)

In a slightly modified version, the IPCC (2014, p. 128) defines vulnerability as 'the propensity or predisposition to be adversely affected. Vulnerability encompasses a variety of concepts and elements including sensitivity or susceptibility to harm and lack of capacity to cope and adapt'. It could be understood as the characteristics of a person or group and their situations that influence their capacity to anticipate, cope with, resist and recover from the impact of climate change or extreme hazard events (Wisner et al., 2004). In this book, vulnerability is referred to as those intrinsic characteristics and manifested conditions of humans and non-humans in a social-ecological system, which could constrain their capacity to cope with, manage or adjust to actual or expected climate and its effects.

There are different models that exist today to carry out community-based risk and vulnerability assessments. The Sustainable Livelihood Framework (Chambers and Conway, 1992; DFID, 1999; Schmidt et al., 2005), The Vulnerability and Capacity Analysis Matrix (Anderson and Woodrow, 1998), The UN/ISDR Framework for disaster risk reduction (UN/ISDR, 2004), The Pressure and Release Model (Wisner et al., 2004), Coupled Human-Environmental Systems (Turner et al., 2003) and the BBC Framework (Birkmann, 2007) are some of the prominent models that could guide our analysis. These vulnerability analysis models help us to understand who the vulnerable groups in the specific community are and analyse the root causes and structural processes by which they are affected by climate change. Some of these models also help us to analyse the consequences of these impacts.

There are also other participatory risk and vulnerability assessment models that could enrich the adaptive innovation process. Field manuals such as the Participatory Risk, Capacity and Vulnerability Analysis by ACF International (ACF, 2012), the Vulnerability and Risk Assessment by Oxfam International (Turnbull and Turvill, 2012) and Actionaid International (Undated) help in facilitating a people-centred risk, vulnerability and capacity assessment. Studies have shown that participatory vulnerability approaches such as Oxfam's Risk and Vulnerability assessment models can shift narratives and power dynamics, facilitate listening to and recognition of subjugated voices, strengthen social networks and enable the co-creation of transformative solutions (Morchain and Kelsey, 2016; Morchain et al., 2019). The Vulnerability and Risk Assessment is usually carried out in a workshop mode with diverse stakeholders. The two-to-three-day workshop can be facilitated by the social worker. The participants of the workshop would include representatives from communities, civil society organisations, NGOs, academia, local and national government and the private sector (ibid). The Vulnerability and Risk Assessment has four main steps. The first step is the initial vulnerability assessment. In this step, the participant actors would engage in listing out a series of hazards and issues that are identified and prioritised in relation to the key social groups and livelihood activities in question. This will be followed by the second step, namely the impact chain exercise. During this exercise, the participant actors map the direct and indirect impacts of these hazards and related issues. The third step is the adaptive capacity analysis, which fleshes out ideas for addressing challenges or system inequalities identified during the impact chain exercise. The final step is the aligning findings with opportunities. During this phase, the participant actors would explore possible solutions as well as identify key stakeholders that need to be engaged further (ibid).

From my personal experiences of conducting vulnerability analysis in diverse community settings, I have found that the political economy model of vulnerability analysis, namely the Pressure and Release Model is a much-suited approach for the adaptive innovation model. The Pressure and Release Model views the impacts or outcomes of climate change as the intersection of two major forces, namely those processes generating vulnerability, on the one hand, and the climate and extreme hazard event on the other (Wisner et al., 2004). According to this model, vulnerability can be understood within three progressive levels: root causes, dynamic pressures and unsafe conditions. The root causes are closely linked to the aspect of governance, emphasising the lack of access by vulnerable groups to positions of political power, decision-making structures and resources (ibid). It is also related to prevailing ideologies of dominance, mainly the political and economic systems. The concept of dynamic pressure encompasses all processes and activities that transform and channel the effects of root causes into unsafe conditions, such as rapid urbanisation, deforestation, decline in soil productivity including global warming and climate change (Wisner et al., 2004, p. 54). The adaptive capacities of vulnerable groups can also be affected due to various dynamic pressures. Factors such as lack of access to resources and entitlements and poor control over it could enhance the vulnerability contexts of marginalised actors (Sen, 1999; Wisner et al., 2004). They also face multiple exposures and are negatively affected by dynamic pressures such as climate change, neo-liberal globalisation and top-down approaches to reduce greenhouse gas emissions (Hein and Kunz, 2018). Unsafe conditions imply specific forms in which human vulnerability is revealed and expressed in the temporal and spatial dimension. For example, the inability of specific vulnerable groups to access safe housing facilities could signify their unsafe condition. They will be forced to live in dangerous locations. Such situations can become further complex with the absence of efficient governance mechanisms to deal with such kinds of risk (ibid). The lack of ethical adaptation mechanisms and appropriate risk reduction measures is itself considered as an unsafe condition (ibid).

The political economy framework has been extensively used to analyse extreme hazard events such as droughts, floods, food insecurity, coastal storms and coastal erosion, the spread of infectious diseases and so on. For instance, social scientists have analysed the progression of vulnerability with respect to the recurrent floods in countries such as Bangladesh and India (Wisner et al., 2004). These studies show that the initial pre-hazard conditions of people and their vulnerability are largely generated from the socio-political and economic systems in which they exist. Examining the trajectories of population vulnerability to emerging epidemics in India, Santha and Sunil (2009) have demonstrated that various social, political and economic factors play a key role in increasing the vulnerability of people to emerging epidemics. The framework also demonstrates that the interactions between ecological and social systems are usually complex and non-linear in nature. On the other hand, planned climate change adaptation measures that are prevalent today usually follow a linear course, assuming that one hazard event acts independently of another.

Our role is to enable communities to analyse their own constraints and opportunities in dealing with risk and uncertainties. Participatory vulnerability assessments should focus on various dimensions such as the intersectional demographic composition of vulnerable groups and the root causes of their vulnerability that are manifested in terms of lack of access to resources and decision-making structures (Wisner et al., 2004). Intersectional factors such as class, gender, caste, ethnicity, age, occupation, disability, citizenship status, illness and poor social capital could affect the adaptive capacities of vulnerable groups. The people-centred analysis should also focus on contemporary contextual factors and unsafe conditions that are shaping people's vulnerability at different levels (ibid). Becoming aware of the social, political and economic contradictions that create oppressive situations and naming them is the first step towards bringing about change. Visioning supported by such analytical processes could help actors frame issues in a context-specific and collective manner. It also provides them a reference to compare the current situation with future scenarios and develop suitable action strategies (Boyle et al., 2001). It will also strengthen coordination and communication among diverse community actors.

S.F.F.5.2 Narratives on micro-mobilisation.

I have noticed that different organisations adopt diverse strategies for micro-mobilisation. The United Nations Development Programme (UNDP) has initiated context-specific micro-mobilisation strategies in their different project locations. For example, they have relied on farmer-trainers to mobilise farming communities in the Comoros Islands Archipelago of Africa (Abdallah et al., 2018). The capacity of these trainers was first enhanced through advanced training in different agricultural and animal husbandry aspects. Followed by which, these trainers were involved in mobilising farming groups. I have come across many instances where the government and NGOs have taken the help of theatre artists to mobilise communities. Be it decentralised planning experiments, literacy campaigns or rainwater harvesting, street theatre has been a common mobilisation medium to convey important messages and also organise communities. Often the theatre artists themselves may not be aware of the problem that is being addressed. Therefore, it is our responsibility to make them scientifically aware of and sensitive to the situation first. The artists themselves need to be recognised as ambassadors for the cause. Prior to the initiation of the theatre, these artists can be exposed to both the problem and possible pathways through screening of documentaries, slide shows, group discussions and exposure visits to model sites. It is ideal, if we could identify artists from within communities. However, this may not be always feasible. The artists should further aim at developing a clear plan of action instead of random performances.

We need to be aware of the non-linearity and struggles that practitioners have to face during the micro-mobilisation phase. When I was working with the Tree Growers' Cooperative Project, I was vested with the responsibility of organising village communities surrounding a particular forest range called Yerakonda (Red Mountain). These communities were largely affected by chronic drought, unproductive land, unemployment, distress sales of crops and livestock and poor infrastructural facilities. Most of the vulnerable groups who were dependent on the forests included the landless, small and marginal farmers, pastoralists, hunter-gatherers, women, elderly and children. It was envisaged that these communities could be mobilised as Cooperatives, which could steer the ecological restoration initiatives in the region. In the initial days, people never cared to attend our meetings. We used to persistently continue our village visits and strive to build rapport and trust with the people. It almost took me a year's time to mobilise the communities of the first village to form a cooperative. Within the next few months, the villagers also initiated the watershed development works. However, seeing the initiatives of the first village, thirteen more villages came forward swiftly to form cooperatives and commence watershed development works. In another year's time, all these fourteen communities came together to form a collective federation of people. This federation emerged as a solidarity network, bypassing gender and caste inequalities and began to collectively explore solutions to their larger social and ecological problems. On the other hand, I was unable to mobilise three to four other neighbouring communities surrounding the mountain. They were reluctant to participate in such projects. Diverse contextual factors such as cultural norms, past experiences with outsiders and internal village-level conflicts could have influenced their decisions.

Phase 3: Dialogic ideation

The third phase in the adaptive innovation cycle is dialogic ideation. This phase focuses on actors' ability to imagine diverse pathways of change. These imaginations will be represented through unique narratives of community actors.

Dialogic ideation is a process of strengthening people's ability to engage with and make sense of their interactions to facilitate the emergence of several constructive ideas, deliberate on them, develop new understanding and craft collaborative strategies for the future (Gilpil-Jackson, 2015; Roehrig et al., 2015; Southern, 2015). Innovation platforms should evolve as continuous learning spaces that represent and recognise these narratives that would also nurture a culture of shared values, aspirations and ideas (Southern, 2015). These narratives may be ambiguous, competing, already familiar or entirely new ones and they need to be heard and recognised (Swart, 2015). All concerned actors should be encouraged to listen to the multiple narratives as they are told and heard, without analysing and discriminating them based on their utility. Each one of the participating actors thus has to navigate through a process of synthesis in which they generate reflective insights that can lead to solutions or opportunities for change (Brown and Wyatt, 2010).

To begin with, we can initiate the dialogue with small groups of community actors associated with the innovation platform. Our immediate focus should be to engage with those actors who care to develop suitable adaptation strategies and gradually engage with larger groups and structures. New ideas are most likely to get enacted when actors truly concerned with the situation own up the ideas and are encouraged to act up on them (Bushe and Marshak, 2015). Taking one concern at a time, the innovation platform may explore a range of ideas towards addressing it. Each idea can be presented as visual narratives, such that all participating actors are able to comprehend it (Brown and Wyatt, 2010). These ideas can be prioritised based on the immediate need, local and scientific knowledge and contextual understanding of the issues at hand.

Actors are both enabled and constrained by the narratives and discourses that they sustain through their everyday conversations (Goppelt and Ray, 2015). Dialogic ideation therefore need not be always designed as a structured process. Instead, it could evolve as an integral part of everyday social encounters and shared conversations. These conversations about the present and the future are closely linked to the actors' everyday lived experiences. Best ideas emerge when actors are provided with ample opportunities for joint meaning making to process and transition from their past (Gilpil-Jackson, 2015). We have to convene, join and stimulate conversations in which the meaning of actors' imagination is explored and strengthened in diverse ways (Shaw, 2015). The innovation platforms should emerge as spaces that could nurture a joint exploration of shared possibilities, and free and complete expression of ideas (Swart, 2015). At the same time, we have to be aware of the fragility and unpredictability of transforming some of these ideas into human action (Shaw, 2015).

To develop a more inclusive, empathetic and reflective frame of action, all involved actors should be willing to reimagine their taken-for-granted frames (Gilpil-Jackson, 2015). Stimulating dialogic relationships and shared perspectives rather than focusing on content could help actors to relook their mental frames (Corrigan, 2015). Our role as conversation weavers could facilitate participant actors to proactively make meaning of the situation and continuously engage in dialogic practice (Goppelt and Ray, 2015). Such an approach to dialogic practice has to recognise the needs of multiple community actors, giving a greater emphasis to previously marginalised voices and willingness to examine divergent views. The aim of dialogic ideation is thus to imagine new alternatives that are inclusive and capable of bringing out the best in community actors, transcending communication barriers. The innovation platforms should emerge as inclusive spaces that would nurture spontaneous and useful discussions on certain ideas and find ways to support and keep building on them (Roehrig et al., 2015).

Listening to diverse narratives and being critically empathetic to them is a skill required for facilitating adaptive innovation. The emphasis is on the meaningfulness of how diverse actors engage, contest and negotiate with certain assumptions and ideas that are embedded in their lived experiences (Shaw, 2015). Such dialogic processes will be characterised by actors' sense of urgency, commitment, wonder and discovery (Southern, 2015). The dialogic ideation processes should be facilitated in such a way that there should be deliberations on possible adaptation options to address certain specific sets of climatic risks and uncertainties. Prioritisation and emphasis should be given to those ideas that have the potential to cover a maximum spread of vulnerable groups. Adaptation options should be thus weighed in terms of its significant breadth and depth, which is able to cater to the needs of a large number of vulnerable and marginalised groups. As mentioned earlier, the vulnerability contexts and capacities of these actors need to be taken into account while ideating on adaptation strategies. While local community actors make adaptation decisions and select strategies, they tend to prefer to adopt systems and practices that have been proven successful to them or their peers. Often, exceptionally new practices tend to expose them to new forms of risks, and this perception is fundamental in they are being cautious to innovations (Beckford, 2002). Risk minimisation in terms of livelihood outcomes often influences the decisions of local community actors while strategizing adaptation (ibid). Adaptation options that are indigenous to the contexts of the local community actors should be encouraged, and there should be minimal external dependency. We also have the responsibility to sustain the motivation of actors by not allowing dominant players to kill ideas or solutions that have already received broad involvement and acceptance (Roehrig et al., 2015).

Innovation is stimulated when multiple actors interact and share their ideas, knowledge and opinions to come up with new solutions (Posthumus and Wongtschowski, 2014). Dialogic ideation expedites diverse community actors to not only express their concern and understanding of the situation, but also imagine pathways for future decision-making and action. Dialogic ideation is about collaborative problem solving and deliberating on the best, practical, feasible and often frugal ideas. It is envisaged that in such a collaborative approach, diverse actors would consistently engage in the co-creation of distinct ideas. The focus of ideation should not be simply focussed on the transfer of ideas and information. Instead, it should consist of all possible opportunities to look at the 'how' and 'why' of current situation, where each actor participates in decision-making in a voluntary and self-determined way (Colucci-Gray and Camino, 2016). Multiple solutions could emerge for the problem in focus and each one of these has to be deliberated upon, as such a process could provide more scope to listen to different actors' voices and ideas.

Ideas for adaptation can represent diverse domains such as ecological restoration (Hettinger, 2012; Sandler, 2012), asset conservation, storage and enhancement (Santha et al., 2017), structural adaptation (Santha et al., 2015), livelihood diversification and communal pooling (Agrawal, 2008) and social protection (Jaswal et al., 2015). Through dialogic processes, innovation platforms should evolve as spaces of possibility for experimentation and knowledge creation, information exchange and the spread of good ideas and practice (Bailie et al., 2018). On certain occasions, there will be a need to ideate in an ever-changing environment including climate, resources and the market. Innovation in adaptation can therefore also be about exploring ideas that are new to certain contexts of practice. These ideas can be a new way of irrigating a field or organising women farmer to connect across an alternative value chain, an inclusive policy that supports smallholders in getting bank loans or a combination of all these.

Dialogic ideation is a phase that extends itself from thinking to doing. All participant actors in the innovation platform have to gradually move beyond collective imagining to collective feeling and experiencing of the vision. This is where we step into the action framing phase. Action framing could help community actors to confront the complexities in adaptation through facilitating a process of deeper exploration, connection and learning (Southern, 2015). It is about how participant actors feel and believe to undertake the doing. The reflective processes of thinking–feeling–believing–doing gain momentum through dialogic ideation. Depending on the ideas and imaginations that have evolved, participant actors need to explore what happens when these are put into action. And also examine what additional resources and skills are required to improvise these ideas. That is where the dialogic ideation stage transitions into the action framing phase. This transition will also help actors to focus on the things they have to immediately respond to.

S.F.F.5.3 Narratives on dialogic ideation.

Dialogic ideation is an enriching and dynamic learning process. As mentioned earlier, the ideas have to be evolved based on the knowledge and needs of local community actors. It should never be external driven or imposed from outside. I once happened to be part of a deliberation between animal husbandry experts and livestock owners during a cattle fair. The experts were trying to propagate the benefit of having hybrid cows in drought-prone and water-scarce regions. They were trying to sell their idea basically from the lens of market-based demand and supply. However, the livestock farmers had their own unique understanding and preferred native or indigenous cattle breeds to hybrid imported and cross-bred varieties. They believed that the native breeds had greater ability to withstand severe summer seasons. They also observed that fodder scarcity was a serious issue during summers. Usually the hybrid varieties would require fresh fodder, which will be scarce during summer seasons. On the other hand, the native breeds of cattle could survive with dry fodder. The experts tried to counter their argument in terms of the volume of milk production. However, farmers opined that the milk production among hybrid cows decreased considerably during summers, while it was low but constant for the native breeds.

I once attended dialogic ideation sessions organised by a Gram Panchayat (local governing body) in Kerala, India. These sessions were part of the decentralised planning project that the government had undertaken. To begin with, they held bimonthly meetings with different stakeholders in each ward of the Panchayat. These meetings were largely conducted in spaces where local community members could attend, speak and listen to diverse viewpoints. Followed by these discussions, almost after three months of deliberations, situational analysis and micro-mobilisation, a public forum was organised in a nearby public school. I experienced the forum as very informative and interesting. There were seminar sessions where people from the community and neighbouring regions presented their experiences and ideas. There were experts present, but they often acted as moderators. Simultaneously, an exhibition was also organised showcasing the needs and possibilities across diverse sectors ranging from agriculture, fisheries, animal husbandry, small and microenterprises, alternative energy, health, education and the environment. Boxes were kept for community members to drop in their ideas. Followed by which, the Gram Panchayat began to identify and prioritise planning measures after deliberating on the feasibility of each idea with sector experts.

It is very interesting and a learning experience to see how women collectives participate in the ideation processes. Often these discussions happen in informal spaces, as part of their everyday livelihood practices. Once I met a group of women farmers who were deliberating among themselves on what crops to cultivate and why? These women were from drought-prone villages and were not happy with the cash-crop-driven agriculture in the region. They wanted to make a change. Of course, they were mobilised by an NGO that works in the organic farming sector. Nevertheless, once they decided to choose an alternative pathway, the decision-making spaces and choices were completely theirs. The day I met this group, they were ideating over crop diversification as an alternative strategy, instead of one single crop of groundnut. They preferred to cultivate a traditional variety of millet. Some women among them also suggested that they should go for mixed cropping, with a combination of native millets, pulses and oilseeds. According to them, mixed cropping enabled them to pool risks of crop loss, while the cultivation of millets also ensured that their cattle will get enough fodder. Later, the project team followed up with them on their planned strategy using a seasonal calendar. In the seasonal calendar, they displayed the different months they will prefer to sow and harvest pulses, millets and oilseeds, respectively. They preferred to go for pulses first, as they believed it would improve soil fertility and the crops that follow would benefit from the same.

Phase 4: Action framing

Ideas do not naturally translate themselves into direct material modes of doing. Instead, they often happen through interactive and interpretive processes and depends on the meanings generated by the respective actors' objects of orientation (Goffman, 1974; Snow, 2004). Action framing is a collectively constructed process by which inner experiences and imaginations of community actors are visualised and appropriate meanings are generated for practice. The aim is not only to assign meanings to relevant ideas, events or situations, but also to facilitate each actor to enable oneself and mobilise others to act upon it (Snow and Benford, 1988). This phase of adaptive innovation explores how suitable working models can be developed out of the ideas generated through dialogic ideation. In this phase, working models can be represented through figures, narratives and stories, prototypes or any other sensory experiences in a temporal and spatial plane. The working models can themselves act as vehicles to stimulate dialogic conversations and reflective practice (Collins, 2017). These models should be capable of generating debate among participant actors on the socio-political and ecological implications of their innovation.

Owing to various contextual factors, all ideas may not evolve as actionable projects. The acceptance of ideas and transforming them into material manifestations largely depend on the ability of community actors to act upon it (Snow, 2004). A dialogic frame on what ideas may work and how, or what ideas will not work and the possible reasons for the same could be explored in this context. Based on the outcomes of such dialogic processes, community actors could redesign their working models or sometimes, even look out for newer ideas. During this process, one needs to consistently look out for intended/unintended and desirable/undesirable consequences of ideas when located in practice. The working models should deepen and widen the dialogic processes as well as refine each actor's understanding of the risk and its consequences. Action framing mediated through working models is more about making sense of the future by exploring the dynamics and transformations in the social-ecological system by improvising one's own contexts of practice, and subsequently applying suitable ideas to facilitate that change (Hillgren et al., 2011; Sanders, 2013; Sangiorgi, 2009). The working models should communicate effectively the issue pertaining to the operationalisation of the generated ideas, stimulate discussion and dialogically engage towards developing actionable strategies (Kimbell and Bailey, 2017). There can be more than one working model for a single idea. There must be considerable opportunity and space to compare these working models and decide on the best possible solution. Such an approach will also help in supporting and triangulating the multiple, located and partial perspectives of actors (Suchman, 2003; Westerlund and Wetter-Edman, 2017).

Action framing is not something extraordinary or new for local community actors. Being attentive to the nature of ideas, interventions and their consequences is part of actors' everyday lived experiences. These actors in their everyday life are consistently engaged in observing, reflecting and deliberating on their emergent situations and seek different pathways to address them. Though tacitly, experimentation with raw working models are part of their daily routines. It helps them to feel and see their idea. They rely on an iterative process in which the key ideas are put into practice through small trials or experiments. Such an approach allows them to make mistakes without incurring huge costs, and immediately reflect on their root causes and possible alternatives. Such strategies will also enable them to recognise and anticipate both drivers and barriers to implement new ideas and ways of working (Kendall and Kendall, 1993). They continuously keep on evaluating and eliminating potential solutions and strive to build on those ideas that would work.

Action framing is not about developing some perfect working model. Instead, it could often be a tangible outcome, which could convey a particular idea and its actionable possibilities. During the process of action framing, we should be alert to actors' behaviour that they are not overwhelmed by the idea and its representation. It is also important that they do not fall in love with the idea too early or have unrealistic expectation of the outcome. They should be open and ready to be challenged and hear about the good and bad of an idea. They also need to have the understanding that their specific ideas might not always deliver the desired outcomes. It can also sometimes turn out to be a quick-and-dirty iterative process (Noyes, 2018). Moreover, each of the participant actor should be receptive to others' experience with the working model (Roehrig et al., 2015; Tudhope et al., 2000).

While reflecting on their working models, it is very important that we not only give attention to what is present, but also look at those representations (and whose) that are absent (Westerlund and Wetter-Edman, 2017). This absence of representation is a kind of othering, where the absence is not acknowledged and something is repressed for one reason or another (Law, 2003; Westerlund and Wetter-Edman, 2017). The representation of a single working model seldom does reveal the socio-political contexts of each social encounter. We may have to therefore pursue multiple approaches to develop working models (Westerlund and Wetter-Edman, 2017). Engaging and triangulating with multiple working models would help participant actors to engage in a dialogic process expressing and exploring the diverse, located and partial perspectives of applying their ideas into practice (ibid). Moreover, in order to ensure the inclusion of all possible representations, action framing should involve those subjugated actors, whose voices and active involvement in local development had not been considered before (Novak et al., 2018). Such approaches also aid the co-creation of knowledge that is practical, inclusive and future oriented (Arrigoni, 2016).

During action framing, the participant actors will be in a constant process of reassessing and reappraising their ideas with respect to their practice contexts. While some will show lots of optimism, there will be others who could end up being frustrated with their efforts (Roehrig et al., 2015). They may raise several unexpected questions, place alternative ideas or tell stories of their previous failures. Giving specific attention to these expressions will enable us to understand the situatedness and struggles of each implicated actor (Novak et al., 2018). Dialogic and shared conversations among actors would help in reflectively engaging with these expressions and pursue a design-by-doing approach (Gilpil-Jackson, 2015; Jernsand et al., 2015; Vink and Oertzen, 2018). In this regard, creating a collaborative dialogue between participant actors and the working models is equally important (Schrage, 2000). A constructive indicator of a good working model is when people start giving feedback on smaller details as well. This indicates that the community actors can move towards piloting the model as they are reaching a refinement point (Collins, 2017).

Institutional building and strengthening are also important components of action framing. Enhancing the capacities of the innovation platform by engaging actors to plan and design its day-to-day operations is a key pathway for institutional building. Norms pertaining to the day-to-day operations of the platform have to be evolved in a participatory manner (Crawford and Ostrom, 1995). Most of these norms, if required, can be formalised through a well-debated and agreed-upon written bylaw. The platform can seek a formal institutional identity such as a self-help group, mutually aided cooperative society, federation, private social enterprise, trade union or any other viable organisational form suiting the context. The roles and responsibilities of all concerned community actors and other stakeholders in the innovation platform have to be clearly defined (Pound and Conroy, 2017; Swaans et al., 2014). These roles and responsibilities can be reflected and revised on a periodical basis (Posthumus and Wongtschowski, 2014).

S.F.F.5.4 The story of Hiware Bazaar.

In the 1980s, Hiware Bazaar was one of the most drought-affected village in Maharashtra, India. Lack of rains, poor ability of the soil to retain moisture, depleted water tables, overgrazing and resultant fodder scarcity had worsened the livelihood uncertainties and vulnerabilities of the local community (Anand, 2004; Fernandes, 2007; Gandhi, 2018). People began to migrate extensively to nearby towns and cities in search of work. In 1990, Mr. Popatrao Pawar returned to his native village after completing his post-graduation. He was elected as the President of the local governing body, and under his leadership the village developed an alternate blueprint to deal with drought. Initial afforestation attempts ended up as a failure as the samplings planted were destroyed. They then set to restore their ecological security through enforcing certain norms and '*shramdaan*' (voluntary labour). The Gram Sabha banned the felling of trees, and also planted around 10 lakh trees over a decade. They were able to converge the afforestation and water conservation works with the state's employment guarantee scheme. All such physical works that were undertaken relied on using locally available materials. Cultivating water-intensive crops such as sugar cane and bananas were prohibited. Digging and use of borewells were also banned and drip irrigation was encouraged. Controlled grazing norms were strictly implemented, and fodder collection was capped. Gradually farmers began to adopt the practice of stall-feeding and after a few years they even began to sell large volumes of good-quality milk. The village over the years has developed reliable trust-based networks and has self-organised into a model village. Hiware Bazaar was awarded the '*Adarsh Gaon*' (Model Village) for its efforts in adapting to water scarcity, enhancing ecological security and strengthening alternative livelihoods (ibid).

The existence of a core group that is accountable towards other participant actors make the innovation platform more transparent and trustworthy (Posthumus and Wongtschowski, 2014). Representation of marginalised actors in the core decision-making groups has to be given specific attention. A structure has to be created to regroup actors on a periodical basis until they have built capacities to self-organise (Gilpil-Jackson, 2015). There should be mechanisms to regularly share information and receive feedback on the activities of the platform. Regular check-in and debrief events could help in this regard. The institutional structure should be sensitive to the consequences of their adaptation interventions (Carlsson, 2003). Sensitivity in this context relate to the flexibility of framing norms and rules keeping in mind the complexity and dynamics of the social-ecological system (Boyd and Folke, 2012). Such a knowledge–practice–belief-based understanding could help in shaping appropriate adaptive institutions (Gadgil et al., 1993).

Innovation platforms should gradually emerge as spaces where inquiry, learning and meaning-making happen in a collective environment of trust, respect and creative engagement (Averbuch, 2015; Corrigan, 2015). There should be proper dialogue and planning to deal with the operational costs of the innovation platform as well. The short-term and long-term sources of resource mobilisation have to be deliberated upon. The spirit of volunteerism and self-reliance has to be nurtured and developed in this regard. Other strategies would include mobilising community funds, public funds and corporate social responsibility sources. Nevertheless, any potential donor has to adhere to the values of ethical adaptation even in their day-to-day conduct. Action framing has to be enabled through capacity-building strategies as well. In this regard, it is not only about building capacities of community actors on the scientific know-how, but also towards enabling outside experts and bureaucrats to recognise the value of diverse situated local knowledge systems. Such an approach is very important, as the asymmetrical nature of knowledge transfer in innovation can be addressed through bottom-up solutions that respond to the local situation and the interests and values of the community actors, and where they have control over the processes involved and in its outcomes (Smith et al., 2017). Capacity building should also aim at preparing community actors to advance towards the next phase of adaptive innovation, namely piloting and scaling up.

S.F.F.5.5 Narratives on action framing.

The transition from dialogic ideation to action framing is a crucial and most interesting phase in adaptive innovation. This is the phase where participant actors are able to convert their mental images into visual forms of action. Here, I would like to share a particular incident that happened during one of my fieldwork. As this project was aimed at ecological restoration and protection of forests, the project team decided to supply pressure cookers to all households in the project area at a heavily subsidised cost. The aim was to decrease the pressure on forests by reducing the consumption of fuelwood in the region. We discussed about the terms and conditions to provide pressure cookers as well as the benefits of using pressure cookers to the project beneficiaries. The organisation entered into an agreement with a leading pressure cooker manufacturing company. Accordingly, the company distributed the cookers to us, and we supplied them to the beneficiaries. A few days later, a colleague and I visited a remote village where the cookers were distributed. We sat in the usual meeting place and began to have casual conversations with the villagers present over there. Suddenly, I saw some object flying towards us and it fell on the ground with a loud noise 'thud'! After few seconds, we realised that the object that flew towards us was a pressure cooker. There was an old woman standing in front of us with the lid of the pressure cooker in her hand, while she had hurled the base of the cooker towards us. In a fit of rage, she said, 'What horrible vessel did you people give me? I cannot make coffee in it! ... It is a waste of fuelwood, energy and money!' It was then when we realised that we took for granted that everyone knew how to use pressure cookers. We went back to our team and redesigned the whole strategy. What was missing was an action frame between the idea and the practice. Before the distribution of cookers, we ensured that we organise training sessions in every village where people can get the opportunity to come, explore, sense and gain experience in using a pressure cooker. People had lots of doubts and concerns that we were able to address.

Smokeless chulhas (stove) and biogas plants are considered to be very effective in addressing deforestation in rural and forest fringe areas. Today, there are different designs and products available in market. However, when compared with its actual potential, these innovations have not been widely accepted by the rural consumer. After an initial hype over the idea, people tend to discard such innovations. I explored the reasons behind the same. I discovered that the acceptance and sustainable use of such appropriate technologies depend on the users' opportunities to visually experience the actionable forms of the shared mental images and provide real-time feedback on these products and services. Actually, the users need to participate in the decision-making and designing of the product based on their own lived experiences. In this context, I met a social entrepreneur who had a different approach to design smokeless chulhas. He used to visit villages and request women who are in need of a chulha to participate along with his project team in designing it. The women will have a mental image of how a chulha should be. They will share it with the professional designers, who will instantly draw the initial sketches and show them to the women. Based on the feedbacks, the design will be revised. Then they can together construct a model chulha with clay or any locally available eco-friendly material. Women will be constantly providing feedbacks and the model will be customised accordingly. I found that such models had a better acceptance among users than the conventional models.

Phase 5: Piloting

Piloting is the testing of the derived working models in prolonged real-world usage (Novak et al., 2018). Pilots help us to understand how our ideas and working models work in actual situations of practice through repeated iteration. It help us to test our ideas and experience their impacts before further brainstorming (Gilpil-Jackson, 2015). It is important that all participant actors have arrived at a meaningful consensus on the working models that could be tested, further refined and improvised before we commence the pilot (Innovate Change, 2015). A long-term outcome of piloting would not only be co-creating an environment of shared conversations, but also a sustained temperament to persistently innovate. There can be different forms of pilot projects and they can differ based on the purpose. Some pilots can act as experiments to understand the outcomes of a particular working model. They largely serve as impact pilots testing or measuring the effects of implementing the working model. On some occasions, pilot projects may intend to begin some kind of immediate action and at the same time explore the teething problems of a particular working model and its context of practice. Such process pilots help community actors to explore the practicalities of implementing a specific adaptation solution. Certain pilots can be demonstration oriented, enabling other actors to understand and follow specific pathways to adaptation. Nevertheless, these classifications are largely for analytical purposes, and the boundaries between these categories are often blurred. Moreover, iterative and reflective feedbacks are crucial elements of any piloting process. For, piloting in itself has to be understood as a complex iterative process.

Some of the pre-requisites to facilitate pilot projects are as follows. The pilot project should aim at addressing some aspects of the root causes and dynamic pressures, and not merely the manifested unsafe condition of people's vulnerability. The participation of subjugated actors and other vulnerable groups in the pilot projects has to be ensured and their feedbacks require greater recognition. The pilot project should take into consideration the systemic inter-linkages of the social-ecological system, and the possible impact the intervention could have on the ecosystem services. Piloting can be done in a frugal and prudent manner, mobilising local resources and expertise to the maximum. There needs to be adequate planning for a systematic gathering of evidence and improvising based on feedback (Patton, 2011). In addition, community actors should be equipped to not only appreciate the success of pilots but also learn from the failures.

Once embarked upon, a pilot must be allowed to run its course (Jowell, 2003). All participant actors have to maintain a commitment to continue with the piloting process even if the outcome is delayed due to certain contextual or situation-specific factors. Such commitments have to be generated through people-centred and participatory decision-making processes. These processes would also help in creating a sense of ownership towards the projects. Moreover, engaging those community actors who have the potential to become future implementers and those who represent the impact groups is likely to produce interventions that are relevant, appropriate, feasible and sustainable (WHO and ExpandNet, 2011). During piloting, we need to ensure that external experts and other similar stakeholders do not dominate local community actors in their decision-making and implementation. Pilot projects must be designed in such a way that they represent and recognise the local and practical knowledge systems of participant actors and at the same time are coherent to the national and international level policy frameworks. The pilot projects also provide us an opportunity to understand how the innovation platforms will evolve in action, the emerging power dynamics between various actors, and how effective these platforms could turn out to be in developing appropriate adaptation strategies. The lessons learned during piloting have to be deliberated upon and contextualised while planning for scaling up.

Piloting is not only testing and demonstrating a model but also refining it through an ongoing learning process (WHO and ExpandNet, 2011). It is a continuous and iterative process of reflection-in-practice. The planning and implementation of the pilot should correspond to the shared vision of the innovation platform. At the same time, the community actors should also be ready to revisit their vision and ideas in the context of both intended and unintended consequences that are unravelled through the pilot. Simpler the pilot intervention, easier will be the implementation of the scaling up project. All pilot projects may not deliver the intended outcome. If unsatisfactory in the first pilot, alternative pilots can be initiated to explore pathways towards the desired situation. Premature scaling up is not advisable as it could result in excess wastage of resources and if it fails to meet the aspirations of involved actors, it could also hinder their motivation. However, if there is pressure to scale-up prior to the project's completion, participant actors could explore options of what elements of the pilot can be safely and successfully scaled up before the final results are available (ibid).

Scaling up of pilot projects involves expanding and gradually implementing the innovation at a large spatial and temporal context. Given the complexities in climate change adaptation, scaling up of adaptive innovation projects will always be an extension of the pilots characterised by continuous iteration and reflection. It is a process of institutionalising the small-scale experiments, prototypes and step-by-step changes into the larger social-ecological system (Roehrig et al., 2015). It involves 'deliberate efforts to increase the impact of successfully tested pilot, demonstration or experimental projects to benefit more people and to foster policy and programme development on a lasting basis' (WHO and ExpandNet, 2011, p. 1). Scaling up involves joint exploration of structures and processes that can support the implementation of larger refined projects as well as in sustaining these initiatives till the next wave of change, if any (Roehrig et al., 2015). Feasibility of pilots should also be assessed in terms of the ability to scale-up the adaptation strategy. Commitment to financial support often becomes a crucial element in deciding the successful scaling up of a pilot (WHO and ExpandNet, 2011). Scaling up may therefore also involve public, private or civil sphere-based service systems or a combination thereof. This involves assessing the financial, natural and human resources available; the ability of the scaled-up project to address the needs of a wider segment of marginalised groups; and the scope to iterate and customise the strategies according to the expectations of different categories of community actors.

S.F.F.5.6 Narratives on piloting.

Piloting is a strategic phase of adaptive innovation that leads us towards shaping the feasibility of ideas and working models. The International Crops Research Institute for the Semi-Arid Tropics (ICRISAT) has revealed the diverse benefits of conducting pilots by providing examples from the Kotha-pally watershed project in the Ranga Reddy District of Andhra Pradesh, India. The project adopted an incremental innovation approach and began working with communities towards water and soil management. As the project progressed, it began to explore other ideas such as crop diversification and livestock integration (ICRISAT, 2009). Various other components were also subsequently piloted such as strengthening market linkage, diversifying livelihoods and setting up of self-sufficient drinking water facilities. A recent innovation of the project was to incorporate weather management controls at local level, such that even children are able to record weather variations (Kane-Potaka, 2016). During this piloting phase, more than 10,000 farmers were trained to work as farmer facilitators. Several institutions such as the Watershed Association, User Groups and Self-Help Groups evolved as innovation platforms for this project. Today, this pilot has been scaled to 28 other regions of the country.

In a similar vein, the UNDP has partnered with the communities in the Solomon Islands in adapting to issues such as drought, increase in salinity of wells, seawater intrusion and subsequent crop loss. Most of the innovations were based on local knowledge systems and resources available. For instance, community actors designed an innovative approach to deal with increase in salinity and crop loss, by using old worn out canoes (UNDP, 2018). These canoes were filled with bags of rich soil and leaf litter, which became a rich nursery bed for growing kitchen garden crops. These raised canoe garden beds were not affected by salinity levels or the extreme heat; and could be moved in case of too much sun. Such innovations assured the participant households to improve their food security (ibid).

Popularly known as 'God's Own Country', the state of Kerala in India has diverse social innovations to its credit. One such innovation has been its attempt to scale up the construction of roads made out of plastic waste. An initial pilot in this regard occurred when Rajagiri College, a socially conscious educational institution in the state built 500 m of polymerised roads using plastic waste in 2012. Inspired from this pilot and other similar stories, a nearby local governing body, namely the Eraviperoor Gram Panchayat began to use shredded plastic with bitumen or asphalt for tarring roads. Gradually, it was framed as a viable solution to deal with the problem of plastic waste in the state. The state government further scaled up this activity through its various social mission projects. Further in 2017, the government of Kerala launched the *Suchitwa Sagaram* (Clean Seas) mission to collect, clean and shred marine plastic wastes that are offloaded along with the fish catch. As a pilot, few fishing villages were involved in the collection, segregation and shredding of plastic wastes from the sea. Amidst all these innovations, Kerala witnessed a severe deluge in the year 2018. The floods caused by intense rainfall and inevitable opening of dams did result in the submergence of several homes and destruction of roads and infrastructure. Road rebuilding has now become a key priority of the government, and certainly the innovation of using shredded plastics in road building could be further scaled up. Nevertheless, despite the pilots being a success and is popularly acclaimed as a well-intentioned initiative, the innovation may be doing more harm than good. Environmentalists point out that the plastic ingredient under conditions of extreme heat could release highly toxic dioxins and other pollutants into the ecosystem. In the long run, the use of plastics for road making may not be ecologically sustainable.

Phase 6: Emergence

Most of the current adaptation and social innovation models are designed to maintain a stable system that is capable of delivering a predetermined outcome. However, when we deal with complex systems such as climate change, we need to factor in the phenomenon of emergence instead of predetermined linear and stable outcomes. Emergence relates to those novel and creative phenomena that arise from and depend on some more basic phenomena; and yet can remain independent from that base (Bedau and Humphreys, 2008). While some adaptation strategies could have enhanced the adaptive capacities of social-ecological systems, certain other strategies would have failed to do so. There can be instances where the adaptation measures would not have reduced existing vulnerabilities, instead would have reinforced them. This could have happened even after taking appropriate precautionary measures. There could also be the emergence of new social practices that are empowering certain segments of the population or environment (Hauskeller, 2015; Sherwin, 1992).

Emergence is characterised by a dynamic process of creativity, flexibility and novelty (Bedau and Humphreys, 2008; Capra, 2002; Sawyer, 2005). Memberships, visions and purpose of the innovation platforms will change with time as new situations of practice emerge (Tui et al., 2013). New rules of behaviour, values, intentions, power relations and leadership styles could influence further progression of the adaptive innovation process. Subsequently, it could result in the creation of new knowledge, skills and ideas (Capra, 2002). As problems are solved and new issues emerge, the activities and focus of an innovation platform may change over time. In this regard, community actors would have drawn new mental images or visions of the future. In extreme situations, the innovation platforms must be prepared to be dissolved once they have achieved their mandated impacts; and also, be ready to develop new organisational forms, if required (Posthumus and Wongtschowski, 2014). There is no blueprint to suggest how such changes are brought about. The form, activities and changes of innovation platform co-evolve with whatever is happening in the wider context of practice (Jiggins et al., 2016).

The emergence of novelty will be usually preceded by a sense of uncertainty, fear, self-doubt and confusion (Capra, 2002). These emotions could create a wrong resistance to change. We have to always be alert to the tension arising out of the interaction between our designed structures and emergent structures. It requires considerable skill to balance the creativity of emergence with the stability of design (ibid). We may end up struggling to comprehend this new reality; and some of these unanticipated events may not fit our existing understandings as well (Capra, 2002; Patton, 2011; Schön, 1985). We may require new knowledge, ideas and skills to deal with this new reality (Capra, 2002). Strengthening communication networks with multiple feedback loops would help us to gather new insights and apply them to the adaptive innovation process (Capra, 2002; Patton, 2011). Our communication has to be honest and open as well.

Emergence can also witness new forms of conflict arising among diverse actors. Some community actors may persist and continue to participate with a single identified strategy, while some others may shift from one creative process to another. Few community actors would temporarily disengage from such collective process and demonstrate a sense of abeyance. In extreme circumstances, some actors would permanently disengage from all the collective action processes (Corrigall-Brown, 2011). We have to therefore learn not only to explore emergence but also engage with it. Instead of dealing with emergence in a negative sense, we have to look at it in terms of developing positive relationships and building opportunities out of it (Holman, 2010). Building trust and mutual support becomes crucial in this context (Capra, 2002). The emergence of trust and leadership can evolve as a powerful means for people to collectivise and address issues of marginalisation (Bergman and Montgomery, 2017). It could also lead to a self-organising process, where community actors are recognised as creative agents of change.

S.F.F.5.7 Subjugated voices, intended and unintended consequences.

Each story-narrating emergence is quite unique, and the patterns and outcomes are often varied. During the 1990s in India, the Tree Growers' Cooperative Society (TGCS) was considered to be an ideal institutional platform to restore ecologically degraded wastelands through community participation and local governance. However, what the community mobilisers did not realise then was that once the commons such as waste lands or grazing lands scale up and embed into larger networks of production and trade that is beyond the local, the management of the resource may be notionally 'owned' by the local, but the use of it could be determined and appropriated by a distant market. In this context, the local needs and vulnerabilities could often get neglected. So, what started as a project to address the needs of the marginalised could itself displace the vulnerable groups from accessing commons in the name of a larger public good.

This itself is a characteristic of emergence and raises certain ethical dimensions of institutional adaptation as well. Theophilus (2002) provides the example of two TGCSs in India. The initial aim of these TGCSs was to meet their minimum requirements of fuelwood and fodder. Over the years, they were successful in regenerating a good stand of Prosopis juliflora, which has high calorific value and makes excellent fuelwood and charcoal. Within a span of five years, these stands of Prosopis were 'ready' to be harvested; and the cooperatives were approached by charcoal traders who trade with big industries in the city. They were ready to offer a higher price than what prevailed in the market. However, the poor and landless in the village were unable to afford to pay a similar price for fuelwood and neither were they able to raise their voices. This was then construed by dominant community actors as a lack of local 'demand' and thereby labelling the fuelwood as 'surplus'. Subsequently, their harvests were sold to the charcoal traders. The money thus earned by the Cooperative was used to pay for the watch and ward of the area. Five years later, the same decision to harvest and sell to traders continued. However, all the protection and trade mechanisms that evolved gradually displaced the poor and landless from certain encroached cultivable areas and grazing lands in the name of voluntary restraint for the larger public good (ibid). Aspects such as meeting local needs of fodder, fuelwood and small timber and ecological restoration were however neglected in the long run.

Emergence could also result in the convergence of new forms of knowledge relationships. Convergence, in this context signifies the cognitive and developmental nature of innovation, where knowledge is located and represented through novel combinations and new relationships (Wienroth and Rodrigues, 2015). Convergence describes 'transformative processes that lead to the creation of something novel beyond the sum of those aspects that are converging: knowledge, practices, stakeholders, artefacts, or spaces' (Wienroth and Rodrigues, 2015, p. 39). These processes result in the gaining of new values beyond synergism leading to the emergence of completely new ideas and outcomes and can lead to new social formations (Glasner et al., 2006; Robinson, 2015). There is enormous potential for converging people's knowledge, science, policy, technologies and practice to help achieve sustainability goals. For example, convergence could facilitate the development of new sustainable technologies for recycling water, saving energy or monitoring sustainability and adaptation indicators (Diallo et al., 2013; Tonn et al., 2013).

The notion of emergence also implies that certain phenomena are theoretically unexplainable or unpredictable in terms of their meaning (Hempel and Oppenheim, 2008). What is emergent with respect to the theories available today may lose its emergent status tomorrow. Therefore, the phenomena of emergence also indicate the scope of our knowledge at a given time (ibid). In an extreme possibility, emergence can be simply a sign of our ignorance as well (Bedau and Humphreys, 2008). The notion of emergence therefore cannot be considered in an objective way (Pessa, 2002). Different actors will see and experience emergence in different ways. Emergence can be defined only relatively to a given observer; whose features should be specified in a suitable way (ibid). This also implies that the subjective experiences of actors have to be an integral part of our understanding (Capra, 2002). The thrust to incorporate emergence in designing adaptation is to be alert to actors' creation of their own meanings and uses for a particular system (Dourish, 2004; Vogiazou, 2007).

S.F.F.5.8 When the pest becomes a cuisine.

A close look at emergence also helps us to understand how a pest emerged as a cuisine. Climate change induced severe beetle menace in Majuli Island on the Brahmaputra in Assam, India. After 2005, the insect became a menace; infesting crops such as potato, sugarcane and green gram. Farmers and scientists attribute it to climate change, after observing prolonged cycles of temperature rise, erratic rainfall and early onset of summer. All these resulted in an increase in the beetle population. The disruptions in the arrival of the main predator to the beetle, namely the migratory Siberian crane during the winter season has also led to the growth of the beetle population. After many iterations, farmers have now transformed the threat into a culinary delicacy by introducing beetles as a nouvelle cuisine (Borah, 2016; Gandhi, 2018). They have developed various dishes such as roasted beetle fry with tomato, plain roasted beetle and beetle curry. Scientists who collaborated with the farmers in this innovation confirm that the beetles are safe to eat and are rich in proteins and carbohydrates (Borah, 2016).

A focus on emergence is also crucial for generating a reflective inner meaning among actors on their thinking, doing and being (Capra, 2002). Emergence helps us to remember the importance of voluntariness, persistence, self-determination and self-awareness of participant actors in the adaptation process (Vogiazou, 2007). It could also result in the generation of new meanings that evolves constantly through shared conversations, observations and actions (Streatfield, 2001). There should be adequate time for self-reflection and to organise, network and have shared conversations. Whenever there are opportunities to celebrate success, they have to be recognised and celebrated collectively (Capra, 2002). All emergent solutions may not be always viable. There should be adequate space to make mistakes, where experimentation is encouraged and learning is valued as much as success. The more open, participatory and situated the interventions are, we could observe the outcomes of emergence with more clarity. Emphasising on emergence is also giving up control, giving way for system self-governance and learning from it (Patton, 2011).

Adaptive innovation as a means for reflective practice and social innovation provides meaningful pathways to involve in the phenomena and practice of climate change adaptation. It has the scope to blend understanding with change/ action, where diverse community actors in specific social worlds become the agents of action and reflection. Adaptive innovation is also a medium for personal and professional growth, enabling social workers to develop a better understanding regarding the particulars of a specific practice-based situation. It is certainly a suitable action research-based orientation for social workers to generate a specific epistemology of practice with respect to climate justice and adaptation. The ultimate belief being that if community actors can analyse, design, implement and reflect on their work in a co-creative manner, then there are greater chances of finding effective solutions. Through adaptive innovation, we will get the opportunity to engage in novel and experimental actions with the aim of finding relevant ideas or solutions to climate risks and uncertainties. It is particularly suitable to ensure the cross-fertilisation of knowledge between all community actors in the innovation platform and thus enhance their capacities to explore about real-world climate change adaptation options. The action research frame inherent in the adaptive innovation cycle provides a flexible and co-evolving process facilitating incremental and transformative change.

References

Abdallah, R.A., Lessire, L., and Benchwick, G. (2018). Resilience in the 'perfume islands': adapting agriculture to climate change in Comoros, *Climate Adaptation UNDP*, 26 July 2018. Retrieved from https://undp-adaptation.exposure.co/resilience-in-the-per fume-islands [Last accessed on 18 July 2019].

ACF. (2012). *Participatory risk, capacity and vulnerability analysis: a practitioner manual for field workers*, Paris: Action Contre la Faim. Retrieved from www.preventionweb.net/files/ 34092_34444acf2013practicalmanuelpcva1[1].pdf [Last accessed on 11 September 2019].

Actionaid International. (Undated). *Participatory vulnerability analysis: a step-by-step guide for field staff*, London: Actionaid. Retrieved from www.actionaid.org. uk/sites/default/files/doc_lib/108_1_participatory_vulnerability_analysis_guide.pdf [Last accessed on 18 July 2019].

Adger, W.N., Dessai, S., Goulden, M., Hulme, M., Lorenzoni, I., Nelson, D.R., Naess, L. O., Wolf, J., and Wreford, A. (2009). Are there social limits to adaptation to climate change? *Climatic Change*, 93(3–4), pp. 335–354.

Agrawal, A. (2008). The role of local institutions in adaptation to climate change, Paper presented at the World Bank workshop on the 'Social dimensions of climate change', 5–6 March 2008, Washington, DC.

Alemagi, D., Oben, P.M., and Ertel, J. (2006). Implementing environmental management systems in industries along the Atlantic coast of Cameroon: drivers, benefits and barriers, *Corporate Social Responsibility and Environmental Management*, 13(4), pp. 221–232.

Anand, N. (2004). Hiware Bazaar: community stewardship of water resources, in *Seeds of hope: case studies from Planning Commission and Lokayan*, India Water Portal, pp. 83–90.

Retrieved from www.indiawaterportal.org/articles/seeds-hope-case-studies-planning-commis
sion-and-lokayan [Last accessed on 5 November 2019].

Anderson, M.B., and Woodrow, P.J. (1998). *Rising from the ashes: development strategies in times of disaster*, 3rd edition, London: IT Publications.

Arksey, H., and O'Malley, L. (2005). Scoping studies: towards a methodological framework, *International Journal of Social Research Methodology*, 8(1), pp. 19–32.

Arrigoni, G. (2016). Epistemologies of prototyping: knowing in artistic research, *Digital Creativity*, 27(2), pp. 99–112.

Averbuch, T. (2015). Entering, readiness, and contracting for dialogic organisation development, in G.R. Bushe and R.J. Marshak (Eds.). *Dialogic organisation development: the theory and practice of transformational change*, Oakland, CA: Berrett-Koehler Publishers, pp. 219–244.

Bailie, J., Cunningham, F.C., Bainbridge, R.G., Passey, M.E., Laycock, A.F., Bailie, R.S., Larkins, S.L., Brands, J.S.M., Ramanathan, S., Abimbola, S., and Peiris, D. (2018). Comparing and contrasting 'innovation platforms' with other forms of professional networks for strengthening primary healthcare systems for Indigenous Australians, *BMJ Glob Health*, 3, p. e000683. DOI: 10.1136/bmjgh-2017-000683 [Last accessed on 17 November 2019].

Barkved, L., de Bruin, K., and Romstad, B. (2014). *Mapping of drought vulnerability and risk*. Final report on WP 2.3: Extreme Risks, Vulnerabilities and Community based-Adaptation in India (EVA): A Pilot Study, New Delhi: CIENS-TERI, TERI Press. Retrieved from www.teriin.org/projects/eva/files/Mapping_of_drought_vulnerability_an d_risk.pdf [Last accessed on 1 November 2019].

Barnett, J., Evans, L.S., Gross, C., Kiem, A.S., Kingsford, R.T., Palutikof, J.P., Pickering, C. M., and Smithers, S.G., (2015). From barriers to limits to climate change adaptation: path dependency and the speed of change, *Ecology and Society*, 20(3), p. 5. Retrieved from http://dx.doi.org/10.5751/ES-07698-200305 [Last accessed on 1 November 2019].

Beaulieu, N., Silva, J.S., and Plante, S. (2015). Using a vision of a desired future in climate change adaptation planning: lessons learned in the municipality of Rivière-au-Tonnerre (Québec, Canada), *Climate and Development*. DOI: 10.1080/17565529.2015.1064807 [Last accessed on 17 November 2019].

Beckford, C.L. (2002). Decision–making and innovation among small–scale yam farmers in central Jamaica: a dynamic, pragmatic and adaptive process, *Royal Geographical Society*, 168(3), pp. 248–259.

Bedau, M.A., and Humphreys, P. (2008). Introduction, in M.A. Bedau and P. Humphreys (Eds.). *Emergence: contemporary readings in philosophy and science*, Cambridge, MA: The MIT Press, pp. 1–18.

Beichler, S.A. (2015). Exploring the link between supply and demand of cultural ecosystem services – towards an integrated vulnerability assessment, *International Journal of Biodiversity Science, Ecosystem Services & Management*, 11(3), pp. 250–263.

Bekkers, V.J.J.M., Tummers, L.G., Stuijfzand, B.G., and Voorberg, W. (2013). *Social innovation in the public sector: an integrative framework*, LIPSE Working papers (No. 1), Rotterdam: Erasmus University Rotterdam.

Bergman, C., and Montgomery, N. (2017). *Joyful militancy: building thriving resistance in thriving times*, Chico, CA: AK Press.

Birkmann, J. (2007). Measuring vulnerability to promote disaster-resilient societies: conceptual frameworks and definitions, in J. Birkmann (Ed.). *Measuring vulnerability to natural hazards: towards disaster resilient societies*, Tokyo: United Nations University Press, pp. 9–54.

Borah, A. (2016). The Majuli beetle turns from pest to delicacy, *India Climate Dialogue*, 10 August 2016. Retrieved from https://indiaclimatedialogue.net/2016/08/10/majuli-beetle-turns-pest-delicacy/ [Last accessed on 18 July 2019].

Boyd, E., and Folke, C. (Eds.). (2012). *Adapting institutions: governance, complexity and social–ecological resilience*, Cambridge: Cambridge University Press.

Boyle, M., Kay, J., and Pond, B. (2001). Monitoring in support of policy: an adaptive eco-system approach, in T. Munn (Ed.). *Encyclopedia of global environmental change*, Vol. 4(14), Chichester: John Wiley and Sons, pp. 116–137.

Brown, T., and Wyatt, J. (2010). Design thinking for social innovation, *Stanford Social Innovation Review*. Retrieved from https://ssir.org/articles/entry/design_thinking_for_so cial_innovation [Last accessed on 11 September 2019].

Bushe, G.R., and Marshak, R.J. (2015). Conclusion: the path ahead, in G.R. Bushe and R. J. Marshak (Eds.). *Dialogic organisation development: the theory and practice of trans-formational change*, Oakland, CA: Berrett-Koehler Publishers, pp. 401–412.

Capra, F. (2002). *The hidden connections: integrating the biological, cognitive, and social dimensions of life into a science of sustainability*, New York: Doubleday.

Carlsson, L. (2003). The strategy of the commons: history and property rights in central Sweden, in F. Berkes, J. Colding, and C. Folke (Eds.). *Navigating social-ecological sys-tems: building resilience for complexity and change*, Cambridge: Cambridge University Press, pp. 116–131.

Chambers, R. (1983). *Rural development: putting the last first*, Harlow: Prentice Hall.

Chambers, R. (1994). The origins and practices of participatory rural appraisal, *World Development*, 22(7), pp. 953–969.

Chambers, R. (1997). *Whose reality counts? Putting the first last*, London: Intermediate Technology Publications.

Chambers, R., and Conway, R. (1992). Sustainable rural livelihoods: practical concepts for the 21st century, IDS Discussion Paper, No. 296, pp. 127–130.

Collins, N. (2017). Prototyping for social impact, *Creativity and Design*, 4 April 2017. Retrieved from www.plusacumen.org/journal/prototyping-social-impact [Last accessed on 11 November 2019].

Colucci-Gray, L., and Camino, E. (2016). Looking back and moving sideways: following the Gandhian approach as the underlying thread for a sustainable science and education, *Visions for Sustainability*, 6, pp. 23–44.

Corrigall-Brown, C. (2011). *Patterns of protest: trajectories of participation in social movements*, Stanford, CA: Stanford University Press.

Corrigan, C. (2015). Hosting and holding containers, in G.R. Bushe and R.J. Marshak (Eds.). *Dialogic organisation development: the theory and practice of transformational change*, Oakland, CA: Berrett-Koehler Publishers, pp. 291–304.

Crawford, S.E.S., and Ostrom, E. (1995). A grammar of institutions, *American Political Science Review*, 89, pp. 582–600.

DFID. (1999). *Sustainable livelihood guidance sheets*, London: Department for Inter-national Development (DFID). Retrieved from www.livelihoods.org/info/info_guidance sheets.html [Last accessed on 12 September 2019].

Di Gessa, S. (2008). *Participatory mapping as a tool for empowerment: experiences and lessons learned from the ILC network*, Rome, Italy: International Land Coalition.

Diallo, M., Tonn, B., Alvarez, P., Bardet, P., Chong, K., Feldman, D., Mahajan, R., Scott, N., Urban, R.G., and Yablonovitch, E. (2013). Implications: convergence of knowledge and technology for a sustainable society, in M.C. Roco, W.S. Bainbridge, B. Tonn, and

G. Whitesides (Eds.). *Convergence of knowledge, technology and society: beyond convergence of nano-bio-info-cognitive technologies*, Cham, Switzerland: Springer, pp. 371–431.

Dilling, L., Pizzi, E., Berggren, J., Ravikumar, A., and Andersson, K. (2017). Drivers of adaptation: responses to weather- and climate-related hazards in 60 local governments in the Intermountain Western U.S., *Environment and Planning A*, 49(11), pp. 2628–2648.

Dolleris, C. (2011). *The visioning approach – in community watershed management planning*, Hanoi: CARE. Retrieved from https://careclimatechange.org/wp-content/uploads/2014/09/Visioning_tool.pdf [Last accessed on 12 September 2019].

Dourish, P. (2004). What we talk about when we talk about context, *Personal and Ubiquitous Computing*, 8(1), pp. 19–30.

Eisenack, K., Moser, S.C., Hoffmann, E., Klein, R.J.T., Oberlack, C., Pechan, A., Rotter, M., and Termeer, C.J.A.M. (2014). Explaining and overcoming barriers to climate change adaptation, *Nature Climate Change*, 4, pp. 867–872.

Ekstrom, J.A., Moser, S.C., and Torn, M. (2011). *Barriers to climate change adaptation: a diagnostic framework*, California Energy Commission, Publication Number: CEC-500-2011-004.

Fernandes, A. (2007). Rural transformation through basic technologies, *The Indian Journal of Political Science*, 68(3), pp. 475–482.

Foxon, T.J., Gross, R., Chase, A., Howes, J., Arnall, A., and Anderson, D. (2005). UK innovation systems for new and renewable energy technologies: drivers, barriers and systems failures, *Energy Policy*, 33(16), pp. 2123–2137.

Gadgil, M., Berkes, F., and Folke, C. (1993). Indigenous knowledge for biodiversity conservation, *Ambio*, 22, pp. 151–156.

Gandhi, F.V. (2018). *A rural manifesto: realising India's future through her villages*, New Delhi: Rupa Publications.

Gawel, E., Heuson, C., and Lehmann, P. (2012). *Efficient public adaptation to climate change: an investigation of drivers and barriers from a public choice perspective*, UFZ Discussion Paper 14/2012, Leipzig: Helmholtz-Zentrum für Umweltforschung GmbH – UFZ.

Gilpil-Jackson, Y. (2015). Transformative learning during dialogic organisation development, in G.R. Bushe and R.J. Marshak (Eds.). *Dialogic organisation development: the theory and practice of transformational change*, Oakland, CA: Berrett-Koehler Publishers, pp. 245–267.

Glasner, P., Atkinson, P., and Greenslade, H. (2006). *New genetics, new social formations*, Abingdon: Routledge.

Goffman, E. (1974). *Frame analysis*, New York: Harper Colphon.

Goppelt, J., and Ray, K.W. (2015). Dialogic process consultation: working live, in G. R. Bushe and R.J. Marshak (Eds.). *Dialogic organisation development: the theory and practice of transformational change*, Oakland, CA: Berrett-Koehler Publishers, pp. 371–390.

Gough, D., Thomas, J., and Oliver, S. (2012). Clarifying differences between review designs and methods, *Systematic Reviews*, 1, pp. 1–9.

Guru, B., and Santha, S.D. (2013). People-centred early warning systems and disaster risk reduction: a scoping study of public participatory geographical information systems (PPGIS) in India, Input paper prepared for the Global Assessment Report on Disaster Risk Reduction 2015, UNISDR.

Hauskeller, C. (2015). Diagonal convergences: genetic testing, governance, and globalization, in M. Wienroth and E. Rodrigues (Eds.), *Knowing new biotechnologies: social aspects of technological convergence*, London: Routledge. eBook, pp. 334–386.

Hein, J., and Kunz, Y. (2018). Adapting in a carbon pool? Politicising climate change at Sumatra's oil palm frontier, in S. Klepp and L. Chavez-Rodriguez (Eds.). *A critical approach to climate change adaptation: discourses, policies and practices*, London: Routledge-Earthscan, pp. 151–167.

Hempel, C., and Oppenheim, P. (2008). On the idea of emergence, in M.A. Bedau and P. Humphreys (Eds.). *Emergence: contemporary readings in philosophy and science*, Cambridge, MA: The MIT Press, pp. 61–67.

Hettinger, N. (2012). Nature restoration as a paradigm for the human relationship with nature, in A. Thompson and J. Bendik-Keymer (Eds.). *Ethical adaptation to climate change: human virtues of the future*, Cambridge, MA: The MIT Press, pp. 27–46.

Hillgren, P., Seravalli, A., and Emilson, A. (2011). Prototyping and infrastructuring in design for social innovation, *CoDesign*, 7(3–4), pp. 169–183.

Holland, B. (2012). Environment as meta-capability: why dignified human life requires a stable climate system, in A. Thompson and J. Bendik-Keymer (Eds.). *Ethical adaptation to climate change: human virtues of the future*, Cambridge, MA: The MIT Press, pp. 145–164.

Holman, P. (2010). *Engaging emergence: turning upheaval into opportunity*, San Francisco, CA: BK Publishers.

ICRISAT. (2009). *Kothapally watershed: a successful example of watershed management based on participation, scientific-backstopping and strategic alliances*, Impact Stories, Hyderabad: The International Crops Research Institute for the Semi-Arid Tropics (ICRISAT).

Innovate Change. (2015). *Tools: prototyping*, Innovate Change. Retrieved from www.innova techange.co.nz/news/2015/7/10/tools-prototyping [Last accessed on 11 November 2019].

IPCC. (2007). Climate change 2007: impacts, adaptation and vulnerability, in M.L. Parry, O.F. Canziani, J.P. Palutikof, P.J. van der Linden, and C.E. Hanson (Eds.). *Contribution of Working Group II to the Fourth Assessment Report of the Intergovernmental Panel on Climate Change*, Cambridge: Cambridge University Press.

IPCC. (2014). Annex II: glossary, in K.J. Mach, S. Planton, and C. von Stechow (Eds.). *Climate Change 2014: Synthesis Report. Contribution of Working Groups I, II and III to the Fifth Assessment Report of the Intergovernmental Panel on Climate Change* [Core Writing Team, R.K. Pachauri and L.A. Meyer (Eds.)]. Geneva: IPCC, pp. 117–130.

Ireland, P., and Thomalla, F. (2011). The role of collective action in enhancing communities' adaptive capacity to environmental risk: an exploration of two case studies from Asia, *PLoS Currents Disasters*, October 26, Edition 1. DOI: 10.1371/currents.RRN1279 [Last accessed on 17 November 2019].

Jalonen, H., and Juntunen, P. (2011). Enabling innovation in complex welfare service systems, *Journal of Service Science and Management*, 4, pp. 401–418.

Jaswal, S., Santha, S.D., Kuruvila, A., Datta, K., Sasidevan, D., and Khan, A. (2015). *Climate change vulnerability and adaptive social protection: innovation and practice among migrant workers in Indian cities*, Asian Cities Climate Resilience Working Paper Series 20, London: IIED.

Jernsand, E.M., Kraff, H., and Mossberg, L. (2015). Tourism experience innovation through design, *Scandinavian Journal of Hospitality and Tourism*, 15(Suppl. 1), pp. 98–119.

Jiggins, J., Hounkonnou, D., Sakyi-Dawson, O., Kossou, D., Traoré, M., Röling, N., and van Huis, A. (2016). Innovation platforms and projects to support smallholder development – experiences from sub-Saharan Africa, *Cahiers Agricultures*, 25, p. 64002. Retrieved from www.cahiersagricultures.fr/articles/cagri/pdf/2016/06/cagri160103.pdf [Last accessed on 2 November 2019].

Jones, L. (2010). *Overcoming social barriers to adaptation*, Background Note, London: Overseas Development Institute (ODI).

Jones, L., Ludi, E., and Levine, S. (2010). *Towards a characterisation of adaptive capacity: a framework for analysing adaptive capacity at the local level*, Background Note, London: Overseas Development Institute (ODI).

Jowell, R. (2003). *Trying it out: the role of 'pilots' in policy making*, Report of a review of government pilots, London: Government Chief Social Researcher's Office.

Kane-Potaka, J. (2016). A watershed moment: dryland farmers adapt to climate change, *India Climate Dialogue*, 19 December 2016. Retrieved from https://indiaclimatedialogue. net/2016/12/19/watershed-moment-dryland-farmers-adapt-climate-change/ [Last accessed on 23 July 2019].

Kendall, J.E., and Kendall, K.E. (1993). Metaphors and methodologies: living beyond the systems machine, *MIS Quarterly*, 17(2), pp. 149–171.

Kimbell, L., and Bailey, J. (2017). Prototyping and the new spirit of policymaking, *CoDesign*, 13(3), pp. 214–226.

Klandermans, B. (1984). Mobilization and participation: socio-psychological expansions of resource mobilization theory, *American Sociological Review*, 49(5), pp. 583–600.

Klein, R.J.T., Midgley, G.F., Preston, B.L., Alam, M., Berkhout, F.G.H., Dow, K., and Shaw, M.R. (2014). Adaptation opportunities, constraints, and limits, in C.B. Field, V.R. Barros, D.J. Dokken, K.J. Mach, M.D. Mastrandrea, T.E. Bilir, M. Chatterjee, K.L. Ebi, Y.O. Estrada, R.C. Genova, B. Girma, E.S. Kissel, A.N. Levy, S. MacCracken, P.R. Mastrandrea, and L.L. White (Eds.). *Climate change 2014: impacts, adaptation and vulnerability. Part A: global and sectoral aspects*, Contribution of Working Group II to the Fifth Assessment Report of the Intergovernmental Panel on Climate Change, Cambridge: Cambridge University Press, pp. 899–943.

Klepp, S., and Chavez-Rodriguez, L. (2018). Governing climate change: the power of adaptation discourses, policies, and practices, in S. Klepp and L. Chavez-Rodriguez (Eds.). *A critical approach to climate change adaptation: discourses, policies and practices*, London: Routledge-Earthscan, pp. 3–34.

Krzywoszynska, A., Buckley, A., Birch, H., Watson, M., Chiles, P., Mawyin, J., Holmes, H., and Gregson, N. (2016). Co-producing energy futures: impacts of participatory modelling, *Building Research and Information*, 44(7), pp. 804–815.

Law, J. (2003). *Making a mess with method*, Lancaster: Centre for Science Studies, Lancaster University. Retrieved from www.lancaster.ac.uk/fass/resources/sociology-online-papers/papers/law-making-a-mess-with-method.pdf [Last accessed on 12 September 2019].

Long, N., and Long, A. (Eds.). (1992). *Battlefields of knowledge: the interlocking of theory and practice in social research and development*, London: Routledge.

MacCallum, D. (2013). Introduction: the institutional space for social innovation, in F. Moulaert, D. MacCallum, A. Mehmood, and A. Hamdouch (Eds.). *The international handbook on social innovation: collective action, social learning and transdisciplinary research*, Cheltenham, UK: Edward Elgar, pp. 343–345.

Martinelli, F. (2012). Social innovation or social exclusion? Innovating social services in the context of retrenching welfare state, in H.W. Franz, J. Hochgerner, and J. Howaldt (Eds.). *Challenge social innovation: potentials for business, social entrepreneurship, welfare and civil society*, Heidelberg: Springer, pp. 169–180.

Mondal, M.A.H., Kamp, L.M., and Pachova, N.I. (2010). Drivers, barriers, and strategies for implementation of renewable energy technologies in rural areas of Bangladesh – an innovation system analysis, *Energy Policy*, 38(8), pp. 4626–4634.

Morchain, D., and Kelsey, F. (2016). *Finding ways together to build resilience: the vulnerability and risk assessment methodology*, Oxford: Oxfam.

Morchain, D., Spear, D., Ziervogel, G., Masundire, H., Angula, M.N., Davies, J., Molefe, C., and Hegga, S. (2019). Building transformative capacity in southern Africa: Surfacing knowledge and challenging structures through participatory Vulnerability and Risk Assessments, *Action Research*, 17(1), pp. 19–41.

Nieto-Romeroa, M., Milcua, A., Leventonb, J., Mikulcaka, F., and Fischer, J. (2016). The role of scenarios in fostering collective action for sustainable development: lessons from central Romania, *Land Use Policy*, 50, pp. 156–168.

Novak, J., Becker, M., Grey, F., and Mondardini, R. (2018). Citizen engagement and collective intelligence for participatory digital social innovation, in S. Hecker, M. Haklay, A. Bowser, Z. Makuch, J. Vogel, and A. Bonn (Eds.). *Citizen science: innovation in open science, society and policy*, London: UCL Press, pp. 124–145.

Noyes, E. (2018). Teaching entrepreneurial action through prototyping: the prototype-it challenge, *Entrepreneurship Education and Pedagogy*, 1(1), pp. 118–134.

Nyikahadzoi, K., Pali, P., Fatunbi, A.O., Olarinde, L.O., Njuki, J., and Adekunle, A.O. (2012). Stakeholder participation in innovation platform and implications for Integrated Agricultural Research for Development (IAR4D), *International Journal of Agriculture and Forestry*, 2(3), pp. 92–100.

Ojha, H.R., Sulaiman, R., Sultana, P., Dahal, K., Thapa, D., Mittal, N., Thompson, P., Bhatta, G.D., Ghimire, L., and Aggarwal, P. (2014). Is South Asian agriculture adapting to climate change? Evidence from the indo-gangetic plains, *Agroecology and Sustainable Food Systems*, 38, pp. 505–531.

Ostrom, E. (1990). *Governing the commons: the evolution of institutions for collective action*, Cambridge: Cambridge University Press.

Oulahen, G., Klein, Y., Mortsch, L., O'Connell, E., and Harford, D. (2018). Barriers and drivers of planning for climate change adaptation across three levels of government in Canada, *Planning Theory & Practice*, 19(3), pp. 405–421.

Patton, M.Q. (2011). *Developmental evaluation: applying complexity concepts to enhance innovation and use*, New York: The Guilford Press.

Pelling, M. (2011). *Adaptation to climate change: from resilience to transformation*, London: Routledge.

Pessa, E. (2002). What is emergence? in G. Minati and E. Pessa (Eds.). *Emergence in complex, cognitive, social and biological systems*, New York: Springer Science + Business Media, pp. 379–382.

Peters, M.D., Godfrey, C.M., Khalil, H., McInerney, P., Parker, D., and Soares, C.B. (2015). Guidance for conducting systematic scoping reviews, *International Journal of Evidenced-Based Healthcare*, 13(3), pp. 141–146.

Peterson, J., Pearce, P.F., Ferguson, L.A., and Langford, C.A. (2017). Understanding scoping reviews: definition, purpose, and process, *Journal of American Association of Nurse Practitioners*, 29, pp. 12–16.

Piccolella, A. (2013). *Adaptation in practice: increasing adaptive capacity through participatory mapping*, Rome, Italy: Environment and Climate Division: IFAD.

Posthumus, H., and Wongtschowski, M. (2014). *Innovation platforms. Note 1. GFRAS good practice notes for extension and advisory services*, Lindau, Switzerland: GFRAS.

Pound, B., and Conroy, C. (2017). The innovation systems approach to agricultural research and development, in S. Snapp and B. Pound (Eds.). *Agricultural systems: agroecology and rural innovation for development*, Amsterdam, The Netherlands: Elsevier and Academic Press, pp. 371–405.

Raj, S., and Bhattacharjee, S. (2017). *Agricultural innovation systems: fostering conver-gence for extension*, MANAGE Bulletin 2 (2017), Hyderabad, India: National Institute of Agricultural Extension Management.

Rendón, O., and Gebhardt, O. (2016). Implementation of climate change adaptation: bar-riers and opportunities to adaptation in case studies, BASE Report. Retrieved from https://base-adaptation.eu/sites/default/files/5.4_BASE_report_web.pdf [Last accessed on 2 November 2019].

Robinson, D.K.R. (2015). Distinguishing the umbrella promise of converging technology from the dynamics of technology convergence, in M. Wienroth and E. Rodrigues (Eds.). *Knowing new biotechnologies: social aspects of technological convergence*, London: Routledge. eBook, pp. 59–102.

Roehrig, M.J., Schwendenwein, J., and Bushe, G.R. (2015). Amplifying change: a three-phase approach to model, nurture, and embed ideas for change, in G.R. Bushe and R. J. Marshak (Eds.). *Dialogic organisation development: the theory and practice of trans-formational change*, Oakland, CA: Berrett-Koehler Publishers, pp. 325–348.

Sanders, E.B.N. (2013). Prototyping for the design spaces of the future, in L. Valentine (Ed.). *Prototype: design and craft in the 21st century*, London: Bloomsbury, pp. 59–74.

Sandler, R. (2012). Global warming and virtues of ecological restoration, in A. Thompson and J. Bendik-Keymer (Eds.). *Ethical adaptation to climate change: human virtues of the future*, Cambridge, MA: The MIT Press, pp. 63–79.

Sangiorgi, D. (2009). Building up a framework for service design research, Paper presented at 8th European Academy of Design Conference, 1–3 April 2009, The Robert Gordon University, Aberdeen, Scotland.

Santha, S.D., Jaswal, S., Sasidevan, D., Datta, K., Kuruvila, A., and Khan, A. (2015). *City adaptation strategies: recognising livelihood struggles of migrant workers in India*, Asian Cities Climate Resilience Working Paper Series 19, London: IIED.

Santha, S.D., Sasidevan, D., Jaswal, S., Khan, A., Datta, K., and Kuruvila, A. (2017). Cli-mate justice, social protection and just adaptation: the vulnerability contexts of migrant workers in Indian cities, in D. Archer, S. Colenbrander, and D. Dodman (Eds.). *Responding to climate change in Asian cities: governance for a more resilient urban future*, Oxon: Routledge-Earthscan, pp. 17–32.

Santha, S.D., and Sunil, B. (2009). A malady amidst chaos: examining population vulner-ability to the chikungunya epidemic in Kerala, India, *Loyola Journal of Social Sciences*, 23(2), pp. 111–129.

Sawyer, R.K. (2005). *Social emergence: societies as complex systems*, Cambridge: Cam-bridge University Press.

Schmidt, A., Bloemertz, L., and Macamo, E. (Eds.). (2005). *Linking poverty reduction and disaster risk management*, Eschborn: Bundesministerium für wirtschaftliche Zusamme-narbeit und Entwicklung.

Schön, D.A. (1985). *The design studio: an exploration of its traditions and potentials*, London: RIBA Publications.

Schrage, M. (2000). *Serious play: how the world's best companies simulate to innovate*, Boston, MA: Harvard Business School Press.

Schusler, T.M., Decker, D.J., and Pfeffer, M.J. (2003). Social learning for collaborative nat-ural resource management, *Society and Natural Resources*, 16(4), pp. 309–326.

Sen, A. (1999). *Development as freedom*, New York: Oxford University Press.

Senge, P.M. (1990). *The fifth discipline: the art and science of the learning organization*, New York: Currency Doubleday.

Senge, P.M., Kleiner, A., Roberts, C., Ross, R.B., and Smith, B.J. (2010). *The fifth discipline fieldbook: strategies and tools for building a learning organisation*, London: Nicholas Brealey Publishing.

Shackleton, S., Ziervogel, G., Sallu, S.M., Gill, T., and Tschakert, P. (2015). Why is socially just climate change adaptation in sub-Saharan Africa so challenging? A review of barriers identified from empirical cases, *Wiley Interdisciplinary Reviews: Climate Change*, 6(3), pp. 321–344.

Shaw, P. (2015). Commentary on dialogic process consultation, in G.R. Bushe and R.J. Marshak (Eds.). *Dialogic organisation development: the theory and practice of transformational change*, Oakland, CA: Berrett-Koehler Publishers, pp. 391–399.

Sherwin, S. (1992). *No longer patient: feminist ethics and health care*, Philadelphia, PA: Temple University Press.

Smith, A., Fressoli, M., Abrol, D., Arond, E., and Ely, A. (2017). *Grassroots innovation movements*, Oxon: Routledge-Earthscan.

Snow, D.A. (2004). Framing processes, ideology, and discursive fields, in D.A. Snow, S.A. Soule, and H. Kriesi (Eds.). *The Blackwell companion to social movements*, Oxford, UK: Blackwell Publishing Ltd., pp. 380–412.

Snow, D.A., and Benford, R.D. (1988). Ideology, frame resonance and participant mobilisation, *International Social Movement Research*, 1, pp. 197–217.

Southern, N. (2015). Framing inquiry: the art of engaging great questions, in G.R. Bushe and R.J. Marshak (Eds.). *Dialogic organisation development: the theory and practice of transformational change*, Oakland, CA: Berrett-Koehler Publishers, pp. 269–289.

Steyaert, P., Barzman, M., Billaud, J., Brives, H., Hubert, B., Ollivier, G., and Roche, B. (2007). The role of knowledge and research in facilitating social learning among stakeholders in natural resources management in the French Atlantic coastal wetlands, *Environmental Science and Policy*, 10, pp. 537–550.

Streatfield, P.J. (2001). *The paradox of control in organisations*, London: Routledge.

Suchman, L. (2003). *Located accountabilities in technology production*, Lancaster: Centre for Science Studies, Lancaster University.

Swaans, K., Boogaard, B., Bendapudi, R., Taye, H., Hendrickx, S., and Klerkx, L. (2014). Operationalizing inclusive innovation: lessons from innovation platforms in livestock value chains in India and Mozambique, *Innovation and Development*, 4(2), pp. 239–257.

Swart, C. (2015). Coaching from a dialogic organisation development paradigm, in G. R. Bushe and R.J. Marshak (Eds.). *Dialogic organisation development: the theory and practice of transformational change*, Oakland, CA: Berrett-Koehler Publishers, pp. 349–370.

Thaler, T., Attems, M., Bonnefond, M., Clarke, D., Gatien-Tournat, A., Gralepois, M., Fournier, M., Murphy, C., Rauter, M., Papathoma-Köhle, M., Servain, S., and Fuchs, S. (2019). Drivers and barriers of adaptation initiatives – how societal transformation affects natural hazard management and risk mitigation in Europe, *Science of the Total Environment*, 650(1), pp. 1073–1082.

Theophilus, E. (2002). Private profits but common costs: experiments with tree growers' cooperatives on village commons, Working Paper No. 4, Foundation for Ecological Security.

Tonn, B., Diallo, M., Savage, N., Scott, N., Alvarez, P., MacDonald, A., Feldman, D., Liarakos, C., and Hochella, M. (2013). Convergence platforms: earth-scale systems, in M.C. Roco, W.S. Bainbridge, B. Tonn, and G. Whitesides (Eds.). *Convergence of knowledge, technology and society: beyond convergence of nano-bio-info-cognitive technologies*, Cham, Switzerland: Springer, pp. 95–137.

Tucker, J., Schut, M., and Klerkx, L. (2013). *Linking action at different levels through innovation platforms*, Innovation Platforms Practice Brief 9, Nairobi, Kenya: International Livestock Research Institute.

Tudhope, D., Beynon-Davies, P., and Mackay, H. (2000). Prototyping praxis: constructing computer systems and building belief, *Human-Computer Interaction*, 15(4), pp. 353–383.

Tui, H.S., Adekunle, A., Lundy, M., Tucker, J., Birachi, E., Schut, M., Klerkx, L., Ballantyne, P.G., Duncan, A.J., Cadilhon, J.J. and Mundy, P. (2013). *What are innovation platforms?* Innovation Platforms Practice Brief 1, Nairobi, Kenya: International Livestock Research Institute.

Turnbull, M., and Turvill, E. (2012). *Participatory capacity and vulnerability analysis: a practitioner's guide*, Oxford: Oxfam.

Turner, B.L., Kasperson, R.E., Matson, P.A., McCarthy, J.J., Corell, R.W., Christensen, L., Eckley, N., Kasperson, J.X., Luers, A., Martello, M.L., Polsky, C., Pulsipher, A., and Schiller, A. (2003). A framework for vulnerability analysis in sustainability science, *Proceedings of the National Academy of Sciences*, 100(14), pp. 8074–8079.

UN/ISDR. (2004). *Living with risk: a global review of disaster reduction initiatives*, International Strategy for Disaster Reduction, 2004 version, Geneva: UN Publications.

UNDP. (2018). *Faces of climate change: Solomon Islanders are bracing for the future on the frontline: UNDP is helping them to prepare*, Climate Adaptation UNDP, 27 July 2018. Retrieved from www.thegef.org/news/faces-climate-change-solomon-islanders-are-bracing-future-frontline [Last accessed on 18 July 2019].

Vanni, F. (2014). *Agriculture and public goods: the role of collective action*, Dordrecht: Springer Science+Business Media Dordrecht.

Vink, J., and Oertzen, A. (2018). Integrating empathy and lived experience through co-creation in service design, Paper presented at ServDes2018 – Service Design Proof of Concept, July 2018, Politecnico di Milano, Milano, Italy.

Viterna, J. (2013). *Women in war: the micro-processes of mobilization in El Salvador*, New York: Oxford University Press.

Vogiazou, Y. (2007). *Design for emergence: collaborative social play with online and location-based media*, Amsterdam: IOS Press.

Ward, M. (2016). Rethinking social movements micromobilization: multi-stage theory and the role of social ties, *Current Sociology*, 64(6), pp. 853–874.

Westerlund, B., and Wetter-Edman, K. (2017). Dealing with wicked problems, in messy contexts, through prototyping, *The Design Journal*, 20(Suppl. 1), pp. S886–S899.

WHO and ExpandNet. (2011). *Beginning with the end in mind: planning pilot projects and other programmatic research for successful scaling up*, Geneva: WHO and ExpandNet.

Wiederkehr, C., Beckmann, M., and Hermans, K. (2018). Environmental change, adaptation strategies and the relevance of migration in sub-Saharan drylands, *Environment Research Letters*, 13, p. 113003. DOI: 10.1088/1748-9326/aae6de [Last accessed on 3 November 2019].

Wienroth, M., and Rodrigues, E. (2015). An introduction to social convergences, in M. Wienroth and E. Rodrigues (Eds.), *Knowing new biotechnologies: social aspects of technological convergence*, London: Routledge. eBook, pp. 33–58.

Wisner, B., Blaikie, P., Cannon, T., and Davis, I. (2004). *At risk: natural hazards, people's vulnerability, and disasters*, 2nd edition, London: Routledge.

Wreford, A., Ignaciuk, A., and Gruère, G. (2017). *Overcoming barriers to the adoption of climate-friendly practices in agriculture*, OECD Food, Agriculture and Fisheries Papers, No. 101, Paris: OECD Publishing.

6   Methods of inquiry and practice

Introduction

Adaptive innovation is a process that is embedded in both inquiry and practice. There are certain unique methods and techniques that could facilitate a dialogic inquiry. These diverse methods of inquiry and practice suitable for each phase of adaptive innovation are discussed in this chapter.

Phase 1: Situational analysis

As mentioned in Chapter 5, situational analysis is the process of understanding the status, conditions, trends and key issues affecting diverse community actors in a given social-ecological context. It was also discussed that situational analysis encompasses the processes of conducting a scoping review, participatory mapping and analysis of drivers and barriers to adaptive innovation. The situational analysis thus commences with a more general, broader question and exploration of a particular situation of practice. This can be done through preliminary field visits and simultaneous review of literature. Apart from these steps, our interaction with key informants during our preliminary field visits could help us to broadly map the situation. During these field visits, we should also explore and probe the social-ecological contexts through certain basic questions that are connected to day-to-day life and livelihood practices of community actors. These interactions during our field visits are crucial for building rapport and trust with key community actors. These interactions can be largely mediated through diverse types of interviews. Face-to-face interviews are highly reflective ways in which participants discuss, share information and co-create a narrative about a particular experience, event or set of issues (Poole and Mauthner, 2014). They are central to facilitating the steps of observation and analysis prior to action. These interviews facilitate the creation of narratives that are local and context specific, and legitimise the experiential knowledge of the community actors (ibid). Interviews in the context of action research provide opportunities for participant actors to describe the situation in their own terms; as well as to explore his or her experience in detail and to reveal the many features of that experience, which could have an effect on the issue investigated (Stringer,

2007). These processes also enable the concerned actors to reflect on the nature of events that concern them (ibid).

Interview in itself is a situation of practice that is not always neutral. It is located historically, politically and culturally as well as mediated by race, caste, ethnicity, class, gender, sexuality, disability and other intersectional identities of community actors (Poole and Mauthner, 2014). Good interviews are those that are able to capture the differences or the heterogeneity of viewpoints, and accordingly throw light on the strengths, conflicts and contradictions of the interviewee's diverse social worlds (Greenwood and Levin, 2007). Interviews are an opportunity for mutual education and do not always intend to gather objective data (Winter and Munn-Giddings, 2001). The aim is not just to gain information from the participants but also to facilitate the knowledge creation process by listening, questioning and reacting in the conversation (Caulkins, 2014). It has to be ensured that some key members of the community are involved in the situational analysis from the beginning. They should be equally aware of the research questions and would have a key role in suggesting how best to collect the kind of actionable data. The interview needs to be done in consultation with these community actors who would also have an important role in the finalisation of questions for the inquiry and should be provided the necessary support to analyse and interpret the data as well (Poole and Mauthner, 2014). The interview experience therefore very much depends on our ability to establish roles and relationships beforehand with the concerned actors.

The interview has to provide all participants with adequate scope to explore and express their experience of the acts, activities, events and issues related to the problem being investigated (Stringer, 2007). The interviewer needs to be empathetic to the interviewing process. A good interviewer has to possess the skills of empathy, the ability to listen and to engage the interviewee through a reflective process (Greenwood and Levin, 2007; Mack et al., 2005). We have to engage with relevant community actors by posing questions in a neutral manner, listening attentively to their responses and asking follow-up questions and probes based on those responses (Mack et al., 2005). The interview questions can focus on events and discussions that take place at different sites of practice such as related to changes happening in the social-ecological system and how community actors make sense of these changes and their impacts. We need to take care that our team do not lead participants according to any of our preconceived notions, nor do we encourage participant actors to provide particular answers by expressing approval or disapproval of what we say (ibid). In this regard, we have to be very careful that we do not impose too many leading questions, as they can be dominated by our own ideas, images and perspectives rather than those of the participants (Stringer, 2007). We have to simultaneously reflect upon how our own positionality will affect the interviews, analysis and the action that can result (Poole and Mauthner, 2014). There are different types of interviews that could help us to carry out a comprehensive situational analysis. These include key informant interviews, conversational interviews, in-depth interviews and cognitive interviews. The different variations of interviews are discussed in the following paragraphs.

The *key informant interviews* are usually conducted at the beginning of the situational analysis. The key informants are usually those community actors who are able to provide crucial information on the situation in focus. They are capable of providing rich understanding on their community history and contemporary transitions with specific observations on the emerging risks and uncertainties. They are beholders of a wealth of information about the nuances of everyday life and represent an efficient source of invaluable cultural information (Fetterman, 2008a). The key informants representing a community are individuals who are knowledgeable about their community, cultural norms and responsibilities, and often help the social worker to navigate from his or her own world to the social worlds of the community actors. Key informants can also be leaders or gatekeepers who may help with access to the community or research site (Caulkins, 2014). They could play a significant role to establish a link between us social workers and the community. The key informants can also be formal actors working with the government or civil society organisations who can provide pertinent information on the present situation and status of adaptation strategies. They may also be the custodians of detailed historical data, photographs, manuscripts, knowledge about past and present power relations. They could help us to develop a community profile and a contextual understanding about the nuances of people's everyday life.

Key informants typically provide information through interviews and informal conversation (Fetterman, 2008a). Often, these interviews are relatively brief, open-ended and informal in nature; and the questions are generally like the grand-tour questions so as to gather a basic understanding of the community and the situation of practice. These interviews are informal also due to our adherence to the principle of collaborative intent, whereby we aim to influence the interviewee to become a collaborator (McNiff et al., 1996). Such strategies also help us to gain a broader understanding of relevant issues in particular social-ecological systems and associated discourses among community actors. The competency of key informants is often measured by length of time they have been in the community, knowledge of community and neighbouring communities or organisations, knowledge about a specific topic and type and degree of interaction with community members (Fetterman, 2008a). It is very important that we establish long-term relationships with key informants to continuously gather and reflect on reliable and insightful information (Fetterman, 2008b).

Conversational interviewing is an approach that is used to enhance our understanding of a situation through informal and shared conversations. Such a strategy could facilitate an interactive environment where participants discuss their day-to-day situations in a less hierarchical environment than formal structured interview settings (Roulston, 2008). In a friendly and informal atmosphere, participants are respected as equal partners and are encouraged to freely share their understanding concerning the situation. In between these conversations, we could pose casual questions to participants about what is going on with respect to a particular situation in focus. Nevertheless, there should be adequate flexibility in allowing topic shifts and questions from participant actors, and treat them with respect, reciprocity, care and intensive listening. Such strategies have scope to not only build trust with participant actors but also understand, respect and remain empathetic to their lived experiences, beliefs and perceptions (ibid).

The aim of *in-depth interviewing* is to be as close as to the interviewee's experience. In-depth interviews can be used to collect data on actors' personal histories, perspectives and experiences (Mack et al., 2005). They are also useful to discover the extent and type of consensus and negotiations happening among community actors concerning key issues in their social and ecological worlds. We usually begin the interview with an open invitation requesting the participant actor to tell his or her story (Fielding, 2003; Hollway and Jefferson, 1997). During the interview process, we could attempt to elicit further narration through open questioning and probing of the actors' situated understanding of the studied events. The questions asked during in-depth interview sessions are intentionally specific towards the interviewee's detailing of key events (Hollway and Jefferson, 1997). During in-depth interviews, the person being interviewed is considered the expert and the interviewer is considered the learner; and at the same time has to be alert to one's theoretical assumptions (Hollway and Jefferson, 1997; Mack et al., 2005). During the interview process, adequate emphasis should be given to attentive listening and periodical note taking. Essential meaning of community actors' lives can be understood only if we listen to them (Kasper, 1994). Care has to be taken that whatever is noted down represents the words, phrases and expressions of the interviewee in its true sense.

Cognitive interviewing encompasses a variety of approaches for eliciting qualitative data on how participants interpret and respond to a wide variety of situations. As a memory-capturing approach, cognitive interview is a technique that will help us to capture the memory of relevant community actors in a coherent and comprehensive manner. Gieselman et al. (1986) describe the process of conducting such a cognitive interview in four steps. This approach can be modified towards understanding climate change adaptation scenarios as well. The first step is to reinstate the context surrounding the event that we are focusing upon. In the context of climate change adaptation, this can be instances like impact of drought on a particular crop, water scarcity during a particular year, flash floods due to cloudburst, emergence of a new pest, forest fires or on migration due to hunger and lack of work. In order to capture how people responded to these situations, cognitive interviewing will be certainly of help. The second step is to encourage participant actors to report everything comprehensively, such that the minutest observation from people's memory is also recaptured. The third step is to encourage participant actors to recall the impacts of climate events and adaptation strategies in diverse orders. Encouraging people to look back into their lives and relook at the order of events that unveiled will help people to explore how certain adaptation strategies could have been different, if the order was reversed, shuffled or combined. Such an enquiry will also help in generating actionable knowledge of what to do, what could be done and what should not be done. The final step is to encourage the participant actor to change his or her perspective and look at the situation from the perspective of other actors in their social world (Fielding, 2003; Gieselman et al., 1986). This helps in generating a shared understanding on the similarities and variations among actors' adaptation strategies. Other methods of inquiry that could help us understand the situation are described below.

Oral history is concerned with how people remember and relate to their past. It helps us towards understanding how community actors associate their past experiences to the present contexts and draw meanings out of them (Veale and Schilling, 2004). It involves the elicitation of personal stories, where participant actors narrate a specific part of their life from their memories. These parts can be certain historical moments pertaining to certain climatic or environmental events, period or spatial location as memorised and shared by participant actors. Such a process of eliciting personal reflections from memory requires considerable time and patience for mutual interaction and facilitating the process of telling and listening to life stories (Liamputtong, 2014a). By definition,

> oral history is primary-source material created in an interview setting with a witness to or a participant in an event or a way of life for the purpose of preserving the information and making it available to others. The term refers both to the process and the interview itself.
>
> (Sommer and Quinlan, 2009, p. 1)

Oral history helps us to understand how individuals and communities have experienced the forces of history (BUIOH, 2016; Veale and Schilling, 2004). It provides information that cannot be obtained from any other regular sources, and it gives voice to ordinary and often marginalised people, whose stories might never have been documented otherwise (Chaitin, 2008). It emphasises on the memories and voices of the less privileged and engages with people of different social positions across generations in locating an in-depth account of their memories, past experiences and reflections to the present context (Adams et al., 2015). Thus, oral history has the potential to reveal a different kind of inclusive history that is characterised by ordinary people's lives (Veale and Schilling, 2004).

Oral histories also add immense value to the process of understanding the present situation of actors in its historicity. Oral history is a repository of folk knowledge, which locates history within and outside the community (Thompson, 1978). Through oral history interviews, individuals recount social, historical, ecological and political experiences and events from their own perspectives (Adams et al., 2015). It gives all the participant actors a sense of belonging to a place or in time (ibid). Oral histories help us understand not just what happened, but how those telling the story understood what happened and what they may now think of it (Sommer and Quinlan, 2009). Memories and personal commentaries collected through in-depth interviews form the core of oral history, from which meaning can be extracted for future action (Ritchie,

2003, 2015). Oral history interviews allow participants to tell their stories on their own terms, so they are able to speak in the way they choose. The whole process helps us understand how participant actors perceive their lived experiences and how they connect with others in society (BUIOH, 2016; Liamputtong, 2014a; Sommer and Quinlan, 2009). Oral histories could help us to understand the trajectories and intensities of socio-economic and ecological transitions. The narratives that emerge through oral history could be used to triangulate data collected through other forms of interview and secondary sources of data such as public records, statistical data, photographs, maps, letters, diaries and other historical materials (BUIOH, 2016). More importantly, oral history aids in deconstructing the prominent notions understood through mainstream history, and many times unearth the subjugated voices that it has tend to ignore (Thompson, 1978). The knowledge thus produced is more personal, experiential, subjective and co-created through the interaction between the interviewee and interviewer (BUIOH, 2016; Veale and Schilling, 2004).

We should be well prepared to initiate an oral history interview. The interview method involves audio-taping one-to-one interviews developed from a general, open-ended prompt supplemented by an interview guide that lists major themes as identified from the review of literature or from other relevant social encounters. An interview session is usually best limited to an hour so as to avoid data saturation and interview fatigue. Multiple sessions at regular intervals can be conducted to complete an oral history interview. The number of sessions would depend on whether we are engaged in conducting a life history or concentrating on a smaller segment of a person's experiences (Ritchie, 2015). The conversations can be recorded in video format as well. Field notes are also a reliable medium to capture the important themes and situational contexts emerging during the interview. After following appropriate measures of taking informed consent and editing the documents ascertaining their confidentiality, the recorded interviews could be transcribed or relayed to the community actors so as to develop actionable strategies. The community actors need to be involved in the analysis and interpretation of the interviews keeping in account the wider social, political, ecological and economic context (Angrosino, 2004; Veale and Schilling, 2004).

There are diverse *participatory learning and action (PLA) techniques* that can help us carry out a situational analysis. Participatory research methods could effectively capture and represent the perceptions, experiences and understanding of diverse community actors that are located within certain temporal, spatial and social contexts (Chambers, 1994, 1997). Often action research projects with communities involve the direct observation and understanding of the life and livelihoods of people as well as the resources that support and sustain these livelihoods. We have to spend a great deal of time in directly observing the dynamics and interactions of the social-ecological system. In this regard, transect walk is an observatory group walk method that has its origins in participatory rural appraisal (PRA) approaches, which helps the diverse community actors and social workers to come together, observe and understand the interaction between different sets of actors and natural resources, resource use patterns, access and power relations in local governance and natural resource management (Narayanasamy, 2009). It depicts a cross-sectional view of actor interfaces in the different agro-ecological landscapes.

Participatory mapping techniques such as *mental models* aim at spatially representing the knowledge of local community actors through a process of collective involvement and dialogue. These maps are found to be useful in facilitating the planning of community-based adaptation strategies to climate change and extreme hazard events (Barkved et al., 2014; Di Gessa, 2008; Piccolella, 2013). The focus of the mapping exercise has to be towards identifying the diverse community actors, their social positions and collectively locating the social and ecological system linkages. We have to give specific attention towards mapping the livelihood contexts of subjugated actors and other vulnerable groups. The mapping processes should also capture the transitions happening in the social-ecological systems (Martínez-Harms and Balvanera, 2012). These include mapping aspects such as major shocks and uncertainties, demographic transitions, local economic trends and institutional changes, individual and collective responses to environmental and economic risks, transitions in asset use and its implications on ecosystem services. The outcomes of participatory mapping exercises are drafted as action-oriented statements, which indicate the possible action steps that need to be taken.

Mental models are cognitive constructs that show how diverse community actors make decisions based on how they perceive their surrounding environment (Elsawah et al., 2015). They are visual tools that allow diverse community actors to collectively depict how they understand drivers and impacts of climatic change (Elsawah et al., 2015; Tschakert and Sagoe, 2009). The first step in developing a mental model is to have a series of brief conversations with community actors on the processes and impacts of climate change (Tschakert and Sagoe, 2009). Followed by which, they are asked to either draw a symbol or write in their local language a concept denoting climate change at the centre of a chart sheet. At the left-hand side of the sheet, participants are then asked to discuss and note down all factors and processes that they believe cause climate change. Any relationship between these drivers is represented through arrows. In a similar vein, the consequences of these changes are discussed and noted down on the right-hand side of the sheet. The relationships between these consequential factors are also represented through arrows. Finally, the participant actors are asked to reflect upon those adaptation strategies that could enable them to cope with the impacts of climatic change (ibid). Mental models are highly subjective of participant actors' ideas, perceptions and beliefs about themselves and their social and ecological worlds. These descriptions are often incomplete and may be inconsistent as well (Rouse and Morris, 1986). Being context-specific and dynamic, participant actors build their mental models based on their perceptions of a particular situation at a certain point of time (ibid). With change in time and contexts, participant actors would consciously or unconsciously relook at their mental models

(Elsawah et al., 2015). Participant actors thus tend to continuously reflect and revise their reasoning attributed to a particular situation. It is important to maintain field notes and research diaries throughout the mapping process.

Yet another visual representation and analytical technique that are similar to mental models are *problem trees*. It refers to the participatory analysis of problems by participant actors and their visual representation, where they relate to the first- and second-level causes and effects of a problem (Chevalier and Buckles, 2008). We have to facilitate the process in such a manner that participant actors are able to identify and deliberate upon a core problem for further analysis and action. The participant actors should be encouraged to identify at least five or six existing factors that are directly responsible for the problem (the first-level causes). After a proper analysis of these causes, the participant actors should determine the second-level causes that are directly responsible for each of the first-level cause (ibid). In a problem tree that is analysed and visually displayed, the tree trunk represents the core problem. The roots and rootlets represent the first- and second-level causes of the problem, respectively. The branches and twigs represent the impact or effects of the problem (ibid).

Other forms of participatory learning and action techniques are *timelines*, *seasonal calendars*, *trend analysis* and *venn diagrams*. A timeline is a chronological list of key events, trends and changes identified, presented and analysed in a participatory manner (Narayanasamy, 2009). The timeline helps us to understand the linkages between the past events and the present situation as well as in illustrating recurring themes (ibid). It helps participant actors to identify those events that have created a certain problem or a situation (Chevalier and Buckles, 2008). The patterns and changes in leaderships, institutions, resource use and access, social problems, family/community history all can be represented through timeline. Seasonal calendar is a diagram drawn collectively to showcase a trend in the main activities, problems and opportunities that community actors have to interface throughout the annual cycle (Narayanasamy, 2009). These can be further categorised into major seasons of the region or can be depicted month wise. The rain calendar is an example of the seasonal calendar. It essentially combines a historical timeline with a seasonal calendar, where the participant actors plot rainfall (little/below normal, average/normal or heavy/above normal rainfall) and temperature (normal, high, cold, very cold) conditions experienced over a period of time (Awuor and Hammill, 2009). Participants involve in describing and interpreting the nature, duration, distribution and effects of rainfall and temperature on their livelihoods as well (ibid). Trend analysis explores the nature of change in temporal dimensions. Emphasis is given to study people's accounts of the past, and how certain significant things have undergone changes over a period of time. Aspects of change in relation to ecological histories, land-use and cropping patterns, customs and practices, population and migration and demographic trends can be gathered and analysed using the trend analysis (Narayanasamy, 2009). *Venn diagrams* enable the participant actors to visually depict and reflect upon the various actor interfaces and situational contexts. A venn diagram is a visual depiction of key formal and informal actors (institutions, organisations and individuals) and their relationship with other local community actors. These diagrams help in understanding and analysing the roles and significance of diverse stakeholders in a locality, levels of communication between these actors, as well as in understanding the impact of external interventions on the day-to-day lives of community actors.

Phase 2: Micro-mobilisation

Micro-mobilisation processes should result in a sense of visioning and consensus building among diverse community actors. These processes should also aim at strengthening actor networks, which could evolve as innovation platforms to deal with climate change and extreme hazard events. In order to develop a blueprint of actors' imagination and future action, a series of *individual and group meetings* and *participatory workshops* with different community actors could be conducted. The initial meetings will be generally oriented towards creating awareness and sensitising potential participant actors about the need for a collective engagement. In the beginning phase, these meetings can be arranged through the community leaders and group leaders or any relevant government agency or civil society organisations. The preliminary discussions will be still to reflect on one's own current situations. Follow-up meetings that are scheduled after consultation with the participant actors should deliberate on evolving a shared vision and commitment to work as a collective. In the due course of these meetings, participant actors should also be able to reflect on their vulnerability contexts and identify the most vulnerable groups. These group meetings can be supplemented by convergent and interactive interviews.

Convergent interviewing is an iterative approach that permits in-depth interviewing, while enabling continuous probing and refinement. It helps members of a collective to gather, analyse and interpret their own experiences, opinions, attitudes, beliefs and knowledge that converge around a small set of interviews (Dick, 1990; Driedger, 2008, 2014). This type of interviewing provides a way of quickly converging on key issues that the action research is focusing upon (Dick, 1990). It combines a structured process and unstructured content with both rigour and flexibility. The rigour arises due to the structured means of developing probe questions. The flexibility is due to the way those probe questions allow responsiveness to the data, along with the continuous refinement of process and content as the interviewing proceeds (ibid). The process of convergent interviewing could be summarised as follows. We would identify a group of participants who are interested to forge a collective to address the climate problem. Each of these participants is then interviewed. Each interview is open-ended to begin with. We would encourage our first respondent to continue talking for perhaps 45 minutes and as it proceeds, detailed probe questions can be developed (Dick, 1990). The interview process is largely guided by a general opening question, which demarcates the boundary for the area of inquiry and ultimately seeks to have participant actors comment on both the positive and the negative aspects of it (Driedger, 2014). Each participant actor is asked to reflect, based on his or her experiences on positive aspects about the phenomenon or issue in question. When further probing on this particular dimension attains saturation, the participant is then asked to reflect on any negative aspects of the same phenomenon. For the second and subsequent interviews with different participants in the collective, we would also ask the participants to reflect on those specific aspects raised during earlier interviews, in case they had not been already mentioned by the latter set of participants. This process continues in a sequential pattern until no new issues are being raised over the entire set of interviews (Dick, 1990; Driedger, 2014). Interview guides and short field notes can be used during this phase to facilitate the discussions of priority issues emerging from the interviews.

Interactive interviewing is a collaborative communication process that takes place with participant actors in small group settings. It helps in drawing an in-depth and intimate understanding of people's experiences through sensitive narratives (Ellis, 2008). Emphasising the communicative and joint sense making that occurs in this form of narrative inquiry, this approach involves the sharing of personal and social experiences of participant actors who tell their stories in the context of an emerging collective relationship (ibid). During the interactive session, each participant actor will get the opportunity to share his or her feeling, insights and story with respect to that particular situation of practice. We social workers will also share our story during the interactive session with the group. The understanding that would thus emerge among all actors will help the collective to develop a shared history and create an environment of willingness to work together. It is anticipated that through such interactive sessions mediated by real-life conversations, trust and relationships will be forged among participant actors. It is a process by which participant actors as a group evolve empathetically to recognise each other's life situations, events and experiences, as well as generate new collaborative insights to deal with the situation (ibid). There are many ways in which the actors would narrate their experience. Our role is to facilitate the group to explore the strengths and limitations of these narratives and counter-narratives as well (Riessman, 2008). Listening to the lived experiences of actors, all concerned community actors in the collective have to recognise the centrality of emerging relationships between them and their experiences across diverse situations of practice. Other methods that could facilitate the micro-mobilisation processes are those methods that could generate visual narratives and shared conversations. These methods include photovoice, community film-making, vignettes, and scenario analysis.

Photovoice is an approach where participants use cameras to document their lives and then collectively analyse their work (Flicker, 2014a). Through the medium of cameras, taking photographs and creating diverse narratives, photovoice provides scope to those community actors to be active participants in voicing their experiences and analysing their varied perspectives through the photographs that they take (Barndt, 2014). It provides opportunity for the participant actors in the co-production of knowledge and in representing their perspectives through photographs and accompanying reflective expressions

(Wilson and Flicker, 2014). The photos and accompanying written and oral narratives become the medium for the community actors to initiate a dialogue with experts and state authorities to bring about intended social change (Flicker, 2014b). In this regard, photographs and related visual narratives are used as a strategy for making transformative change (Wilson and Flicker, 2014).

The three major phases of a photovoice project are preparation, production and use. The preparation phase involves the identification of key community actors to be part of the project. These actors consist of photovoice participants, photographer mentors and facilitators and a target audience of policymakers and/or community leaders (Barndt, 2014). It is important during this phase that all the actors are made aware of the concept and method of photovoice, including ethical concerns and issues of informed consent. The photographic mentors train the participants on how to use cameras. The production phase involves participants taking photos surrounding an initial theme of inquiry. During this phase, participants are encouraged to discuss in groups about their decisions to take certain photographs and also their interpretation of those images. During this process, the facilitator works with the participants to identify one or two significant photos, and then to frame stories around these photos. The participants connect these stories to their own experiences. These narratives are then thematically analysed in a participatory manner, exploring the root causes of the problems highlighted and possible solutions for the same. The photographs serve as emotionally charged metaphors for the participants' perspective on the issue, and the facilitator engages the participants in a dialogue to unveil the meaning of the photographs (Cunningham, 2014). In the final 'use' phase, the photographs and associated visual narratives are shared and deliberated with the actors in the participants' social world, policymakers and other community leaders (Barndt, 2014).

In a similar vein, methods such as *community film-making* aims at involving local community actors in openly exploring and sharing what it is like to live in their specific social-ecological systems (Egmose, 2015). In this process of filming their own lives, local community actors self-reflect on their situations of practice. The first step towards facilitating the community-film-making project is to initiate pilot focus group meetings. Across multiple focus group sessions, participants would meet on a regular basis to discuss what it is like to live in their specific social-ecological context. Simultaneously, these participants will be mentored by professional filmmakers to do their own storyboards and shoot their own films reflecting the issues they wanted to address. Similarly, methods such as *vignettes* can also be used as an effective tool for micro-mobilisation. It comprises a stimulus that selectively portrays elements of reality to which participant actors are invited to respond (Hughes, 2008). Vignettes can take a range of written, audio and visual forms. Written text includes short scenarios and extracts from literature and newspapers. Audio vignettes include spoken narratives, music, songs, and sounds. Photography, painting and line drawing can also be used as visual vignettes. Audio-visual vignettes can include films and live performed acts. Vignettes are commonly embedded within interviews and group discussions (ibid). Certain community-made films themselves can emerge as interesting vignettes. All these visual methods can also be combined to facilitate collective engagement and deliberations.

The aim of *shared vision workshops* is to enable local community actors and other stakeholders to engage in a shared dialogue. The aim is to listen and reflect on how diverse community perspectives could become the base for framing ethical adaptation initiatives. The aims and roles of the shared vision workshops are discussed in the initial community-level meetings. The deliberation process of the shared workshop is primarily organised across 3–4 days. There can be multiple sessions with varied themes. There can be sessions that could facilitate the sharing of local knowledge, experiences and understanding of local community actors to all involved stakeholders. The initial sessions can also be aimed to establish a positive and supportive environment for collective action. These interactive sessions can be supplemented by the screening of relevant documentaries. The follow-up sessions can be thematically designed such that actors are able to involve in participatory dialogue and self-reflections (Egmose, 2015). These thematic sessions have to be flexible, and priority should be given to those ideas that the community actors decide as important to them (ibid). In this regard, the use of mental models could help in strengthening shared and reflexive conversations (Senge, 1990; Senge et al., 2010). The shared vision workshops could end with the co-creation and sharing of community charters envisioning ethical adaptation situations. In this regard, the *scenario analysis* could be applied as a participatory situational analysis technique.

Scenarios are descriptions of how the future would evolve, using current information and assumptions about future trends (Bizikova et al., 2009). A scenario analysis is a process of imagining and reflecting upon potential future events by exploring alternative pathways. Given the complex circumstances, local community actors may not be completely sure of how the climate will change in the future, and they might even have lesser confidence on the suitable adaptation options. However, processes such as participatory scenario analysis could help community actors to mutually engage with one another in dealing with such uncertainties. Participatory scenario development enables us to explore how local community actors anticipate the impacts of climate change and design future adaptation options and development pathways. The first step in participatory scenario development is to collectively decide on the scope of the scenario. Then, through a process of dialogue, local community actors are involved in a process of analysing the present contexts of vulnerability and the key factors contributing to the emergent risks. Followed by which, these participant actors develop different scenarios of the future using white boards or chart papers, stick notes, collages or index cards. These processes also enable participant actors to reflect on suitable visions and adaptation options to deal with climate risks and uncertainties. An additional step could be to review the identified scenarios by reflecting on 'what if' factors that could act as barriers or drivers to these scenarios (ibid). If required, we could avail diverse consensus development techniques during the scenario analysis or shared vision workshop. The *matrix ranking* and *scoring technique* is one such method. We could facilitate community actors to ascertain the problems identified in the order of urgency and prioritise them accordingly. This process also stimulates discussions among people whenever any choices are to be made in finding possible solutions to the problems being prioritised (Narayanasamy, 2009).

Phase 3: Dialogic ideation

The third phase in the adaptive innovation cycle is dialogic ideation. This phase will be characterised by a process of shared imagination where community actors strive towards co-creating future pathways to deal with the risks and uncertainties associated with climate change and its impact. As discussed earlier, *convergent interviewing* provides a way of quickly identifying and linking key issues that community actors are exploring to engage with (Dick, 1990). In a similar process, as discussed in the micro-mobilisation phase, convergent interviews can be used in the dialogic ideation phases as well. Apart from these, focus group discussions could facilitate the process of dialogic ideation to a considerable extent. *Focus groups* aim at generating shared conversations through group interaction on a pre-determined topic of interest or concern (Kitzinger, 1994; Morgan, 1996). It enables community actors to enhance their understanding about a particular phenomenon through focused, shared and intimate conversations. While the focused nature of the group comes through its collective activity, it is the discussion that becomes the source of ideas among focus groups (Kitzinger, 1994). Focus groups are applied in situations where there are diverse perspectives on a particular issue, and we need to generate multiple ideas and possible consensus to co-design future pathways. In focus groups, participants raise queries and attempts to find explanations through shared conversations with each other. It allows community actors to exercise a fair degree of control over their interactions and pursue the discussions in a very relaxed manner (Morgan, 1996). It also enables actors to gather multiple perspectives and make meaning out of these conversations in their own language (Alderfer and Smith, 1982).

Focus groups can be initiated based on the insights that we would have gathered from the situation analysis and micro-mobilisation phase. In a focus group, participants with certain traits of homogeneity (or a common goal) are selected purposively and gathered together to share perspectives and thoughts regarding a pre-defined topic (Logie, 2014). Focus group sessions should be of very short durations extending to a maximum of one hour. Each session should also have no more than 8–10 participants. If required, multiple sessions with the same group can be held over a longer time frame. In focus groups, we could facilitate the group interactions and ensure that the discussions are focused and flowing. It is advisable to record the discussion with the permission of participant actors. We could commence the focus group session by explaining the purpose of the session, and the basis for inviting the specific group for the discussion. We could also briefly share the findings of our situational analysis with the group. We could also ask the group to list out certain expectations that could emerge out of these discussions. The focus group session should aim at helping participant actors to explore opinion and experiences on specific ideas, and mutually clarify them in a relatively short time. As the goal is to listen to a range of actors, the defining element of focus groups is the use of the participants' discussion as a form of idea generation. We should ensure that the group composition is decided based on the community actors' point of view. During the discussions, participant actors should be interested and feel comfortable talking to each other about the particular adaptation strategy and possible ideas that could be co-created through the discussions.

Brainstorming workshops are yet another effective method to elicit meaningful ideas in a collective manner. Brainstorming is a participatory process aimed at stimulating a wide variety of possible ideas or solutions from diverse actors. An idea, problem or its solution is deliberated by participant actors in terms of feasibility and impact (McKernan, 1991). Brainstorming workshops turn out to be effective when there is adequate space for the free flow of ideas. Too much criticism and non-encouragement of actors' imaginations could demotivate actors from participating in the process. To begin with, we should explain the context of the session to the participants very clearly and in a simple language. The initial session should focus on problem analysis, such that the participants should be able to grasp the nature of the problem clearly in their own contexts of understanding. We should make arrangements to record the ideas visually, be it in the form of key points mentioned in a whiteboard, stick notes on a wall or any other viable and accepted option. The follow-up sessions could be organised to discuss the ideas that have evolved from the larger group. We have to facilitate the discussions to select the best three or four ideas from this larger pool. It is ideal during this phase to organise the larger group into thematic working groups, and each of these smaller groups could identify one category of idea for further brainstorming and inquiry. At the end of the brainstorming session, these working groups would present their suggestions and way forward to the larger group for further validation and summation. While facilitating the workshop, we should seek a combination of ideas that could enhance the meaningful participation of actors (ibid).

Yet another technique that could facilitate dialogic ideation is organising interactive session in the format of *Socratic Circles*. This method is well suited to situations when a small group of actors with multiple perspectives and expertise wants to explore a topic or issue via open, honest and meaningful discussion (Busse, 2016). To begin with, participant actors have to be informed of the issue at hand and the purpose of ideation. Some basic ground rules have to be explained to the participants. In a spacious room or under a tree, participants could be organised and seated into two concentric circles. The social worker could set the stage for conversation by sharing a previously determined prompt/challenge question to spark ideation. Only those participants sitting in the inner circle are allowed to speak in the beginning. Instruct the participants at the outer circle to remain silent and participate by listening, watching for insights, similarities or meaning. The conversation should be organic and led by the participant actors themselves. The emerging ideas and their associated meanings can be mapped down using a flip chart. Once the group saturates in their ideation, the outer circle can take the place of the inner circle. We could begin the conversation again using the same challenge question or issues. These discussions can also be mapped down in the flip chart. We could then probe further with participants in both the circles on any new insights or ideas and discuss possible future pathways (ibid). Socratic circles facilitate ideation through alternative narratives or through the storying of participant actors' lives (Swart, 2015).

Dialogue conference is yet another method to facilitate ideation. Dialogue conference acts as a space for all participant actors to create a common ground for their own further collaborative activities (Shotter, 2014). The underlying principle is on equal right to speak, where the local actors involved will decide on the things to which they will engage with and accordingly frame the context in which they will attend to (ibid). The aim is to nurture communicative action through the sharing of actors' lived experiences at the grassroots and shape actionable steps surrounding their local contexts and livelihood spaces; designing local community development projects towards livelihood security, ecological restoration and conflict resolution. These processes of dialogue would also encourage elements of reflection and learning among local community actors (Springett, 2014). The steps in dialogue conference can be explained as follows. To begin with, we need to contact the participants who have volunteered in advance and help them understand the essential features of the whole process. Pre-meeting contacts and preparation of participants are essential elements of the whole process. During the sessions, each participant will share stories based on his or her lived experiences in front of the other participants. Adequate care has to be given so that each speaker is not swayed by the others' lived experiences. We have to encourage participants to actively involve in analysing the variations rather than the similarities in these experiences, and to recognise these variations with an empathetic understanding (Shotter, 2014). We have a crucial role to provide and create a supportive co-learning process and reflexive, trustful environment between the participants (Springett, 2014).

Phase 4: Action framing

Action framing involves translating the emergent ideas into meaningful working models. Community actors need to be provided considerable opportunity to compare these working models and decide on the best possible solution. Some community actors will respond best to stories or enactments, and some others to visual working models (Sanders, 2013). For instance, a *storyboard* showcasing the lived experiences of community actors can help them to explore how each one of them would ideally respond to an adaptation strategy (Jackson, 2015). It will be useful if we choose those methods that are culturally and locally suited and relevant. Some of the folk-art forms prevalent among vulnerable groups can be adapted to suit the context. Such an approach could provide a creative yet accessible means for participant actors to express their views and tell their stories and develop strategies to implement their ideas.

Social simulations such as *games experimentation*, *role plays* and *design charrettes* can also be used for action framing. Games can be designed as reflexive practices that could help actors to analyse their motives, navigate across diverse ideas and develop actionable strategies for specific situations. For this purpose, games can be designed as an explicit and carefully thought out processes of sensing, thinking and doing (Ellen et al., 2015). A simulated game session can be organised for a maximum duration of two hours for a small group of 12 to 20 members. The group has to be assigned a fictional climate adaptation challenge to solve under certain pre-specified criteria of drivers and barriers. Each actor will be allocated a specific role in co-designing a local adaptation measure. Questions related to accountability of action also have to be assigned. These role-play simulations will have fixed rules and constrained outcomes, which are given as general instructions to all actors. Participant actors, as part of an imaginary task group should be encouraged to think out of the box and present their own perspectives on the nature of the problem and how to solve them. All necessary information should be provided in advance to the group and considerable time has to be given to deliberate and reflect up on multiple choices identified by them. Follow-up sessions could be held where new drivers and barriers to adaptation can be introduced and actors are encouraged to look out for alternate adaptation measures (Schenk and Susskind, 2015). Each session has to be followed up by debriefing. The final outcome of these sessions could be one or more actionable solutions that everyone agrees to (ibid).

A design charrette is a workshop represented by an interdisciplinary team of social actors, which aims to quickly develop ideas into actionable designs through exploratory sketches, mental models or collages (Roggema et al., 2015). Design charrettes help to collectively work with community actors and other diverse experts to deal with wicked problems such as climate change. To begin with, we could present the available information to the community actors and other external experts on the social-ecological situation in a concise, understandable and visible manner. Followed by which, each actor could present how they see the problem and how they will solve it. In this regard, it is important to rely on the capacities of actors to tell stories based on their lived experiences as well as on their ability to visualise future scenarios (ibid). The next step is to interchange and replace the functional role of each actor with those of other participant actors. Followed by which, small task groups are created to draw sketches and develop mental models or collages that depict possible actionable strategies. These mental models can be images, assumptions and stories that actors carry in their minds of themselves, other actors and how each one of them shape the way they act in their everyday world (Senge, 1990; Senge et al., 2010). Different task groups can represent the same idea through multiple mental models. Developing suitable mental models require meaningful conversations between actors with considerable elements of openness and merit (ibid). In a similar vein, *collage making* is an arts-based research approach to meaning-making through the juxtaposition of a variety of pictures, artefacts, natural objects, words, phrases, textiles, sounds and stories (Norris, 2008). It is a process of communicating one's ideas and strategies of action by creating a new image from the bricolage of resources collected. The aim is to facilitate the mutual sharing of one's ideas and imagination through collectively assembled images. Aspects such as colour, space and time all play important roles in shaping actor's imaginations. They are meant to evoke disparate meanings in others and strive to communicate on a metaphoric plane (ibid). The larger group will then provide feedback on these diverse action frames (Roggema et al., 2015). A particular task group with specific design expertise could enhance and work on the final designs.

Participatory modelling is also a very effective method to enable community actors to participate in complex decision-making and build appropriate adaptation models (Huntjens et al., 2015). This is a facilitated group exercise, where participant actors evolve different mental models of the problem and actionable steps to address them. Each session can be followed by breakout and debriefing sessions, as per the requirements of the group. As the group exercise progresses, the different mental models could be synthesised into a single larger model. However, this is not a mandatory outcome, as adaptive innovation emphasises on the situatedness of each of these models. Participatory modelling provides spaces for local actors and outside experts from multiple domains to collaborate and work towards strengthening the conceptualised plan of action. These experts may participate at different stages of the adaptive innovation enabling local actors to take effective decisions.

In a similar vein, participatory prototyping is a suitable approach for action framing. It signifies the cyclical, intertwined and iterative relationship between making, enacting and telling (Sanders, 2013). Prototyping is a fundamental design competence, which should be seen as an activity for exploring, proposing and creating knowledge (Westerlund and Wetter-Edman, 2017). In a larger context, prototyping in itself is a cyclical iterative process of translating ideas into action and testing these actions and their outcomes from the perspective of diverse actors (Dunn, 2017). Prototyping enables participant actors to uncover unforeseen implementation challenges and unintended consequences so as to ensure some kind of feasibility and sustainability in the adaptive innovation process. In complex situations such as climate change and extreme hazard events, the cause–effect relationships are not understood in advance and actors are also unclear of what the right solutions will be. Small-scale prototyping helps actors during such occasions to understand what they value most, what works and what does not, as well as to test and refine their ideas rapidly (Noyes, 2018; Roehrig et al., 2015). A participatory prototyping session usually has the following elements. We could begin the session by stating the purpose of the session. The generated ideas will be then summarised and shared by different participants. The anticipated aims and outcomes of creating working models will be then deliberated and agreed upon by all the participants. They would also discuss the options to choose and develop different working models within a particular time frame. Attention has to be given to build working models out of local resources. Once the working models are presented to the larger group, adequate provisions should be there to harness timely feedback on the models. The session should end with debriefing and plans for follow-up action.

Yet another technique that could be used to facilitate co-creation of ideas and further frame actionable designs is the *Collaborative Idea Canvas*. It is a diagram drawn collaboratively by the relevant actors to design, strategise and execute an idea. The approach is more of a collaborative and participatory adaptation of the Business Model Canvas, developed by Osterwalder and Pigneur (2009). A typical idea canvas has eight interdependent sections. Surrounding the focal idea, these interdependent sections are the strategy, resources, participant actors, communication, decision-making, execution and measurements. Each section has necessary space for actors to describe the elements that could affect the designing and implementation of the idea. The questions that guide the deliberations are not rigid questions. Instead, they focus largely on the multiple contexts and values shaping the ideation and its implementation.

Phase 5: Piloting

Piloting involves testing the ideas and working models in practice. Adaptation pilots can be wide and varied across spatial and temporal contexts. Watershed management, afforestation, development of inclusive financial instruments such as micro-insurances, capacity building and documenting traditional knowledge systems are all examples of ongoing adaptation pilots in different parts of the world. There are also pilot adaptation projects that are spread across diverse sectors such as water resources, agriculture, forestry, coastal wetlands, health, energy and infrastructure. These pilot projects aim at addressing several climate change issues through bringing about appropriate alterations in land-use and cropping patterns, reducing water scarcity and wastage, initiating soil and water conservation, developing community-based early warning systems, risk insurances and other risk-integrated social protection programmes.

Piloting should be carried out in conditions resembling local contexts and situations in which it will be scaled-up (WHO and ExpandNet, 2011). Usually the tendency is to experiment the pilot in the land or setting of a big farmer, where there is water availability, land and other facilities. Or else the pilot will have scope for lot of donor-based funds, while no funds would have been allocated for the scaling up. However, this excludes the living conditions and asset-based capacities of the poor and vulnerable groups. Instead, the pilot projects certainly have to be tested in the context of the resource less and vulnerable groups; and simultaneously projects have to be designed on how additional resources can be provided to these groups and cultivate ownership and commitment to the project. Community actors should be aware of the resource constraints and alternative choices available. Any external resources or inputs provided during piloting have to be accounted even for the scaling up phase as well.

All community actors need to participate and commit towards identifying the elements that are crucial for scaling up pilot projects. At this stage, the organisational and technical capacity, and the development of links to input supplies, markets and technical assistance of the innovation platforms have to be strengthened (Pound and Conroy, 2017). Resource mobilisation, both in terms of human, physical and financial resources and in terms of endorsement and support, will be crucial to sustain the process beyond the lifetime of the pilot (Swaans et al., 2014). Nevertheless, it is important to remember that most of the social innovations fail as they negate the creativity and intelligence of implicated local community actors, and instead they tend to come packaged with exogenous participatory processes, encourage scaling up and ignore innovation that is already occurring (Matthews, 2017). Facilitation and management therefore have a critical role to play, often involving a transition from an external-led initiative to a self-organised platform (Ngwenya and Hagmann, 2011). Some of the supportive methods that would help us to understand the lived experiences of actors involved in pilot projects are discussed below.

Storytelling is an interesting method for examining the experiences and impacts of pilot projects. This method is indeed central to action research (Greenwood and Levin, 2007). Storytelling could shape a more inclusive and active political space in facilitating dialogue and action among participant actors (Arendt, 1958; Tassinari et al., 2017). A story generally refers to the ordering of events that infers causal relationships between them. Stories have the potential to create new ideas and establish new ways of imagination (Mead, 2014). Stories give rich insights into the lived experience of actors and facilitate a shared understanding on the possibilities for novel action (Zandee, 2014). It aids experiential knowing (Liamputtong, 2014b). It not only aids the transfer of information but also adds confidence to the participant actors (Johnson and Porter, 2014). Telling stories is an attempt to change one's own life by affecting the lives of others (Mead, 2014). Storytelling is thus an imagined (or reimagined) experience narrated with enough detail and feeling to cause the listener's imagination to experience it as real (ibid). It is a fundamental aspect of human experience, where meaning is conveyed through the act of storytelling.

Storytelling needs to be viewed as a political action, taking place through words, where all actors take part in the discussions and take decisions in a collective manner (Arendt, 1958; Tassinari et al., 2017). It enables participant actors to authenticate their claims to knowledge, to assert their individual and cultural identities, to strengthen social networks, facilitate thought leadership, and to reflect on their own situations, current realities, present struggles and future aspirations (Mead, 2014). It brings together the act of telling and the act of making (Arendt, 1958; Tassinari et al., 2017). Storytelling among vulnerable groups has the potential to generate counter-narratives. These counter-narratives act as symbols of pride in their identity and culture, organise these groups as a collective and could build new strategies of resistance against injustice. In this sense, storytelling offers a source of analysis, consciousness and action

(Ledwith, 2017). We have to work with community actors in facilitating mutual storytelling and retelling the story as the pilot project progresses (Clandinin and Caine, 2008).

There are interesting storytelling approaches such as 'listening out loud' that focuses on listening, telling and retelling (Adams et al., 2015; Pollock, 2006). The process begins with a conversation in which participants learn about each other on several aspects such as their identities, lives and experiences. No recording in any form is carried out during this phase. The emphasis is for each participant to absorb the other person's story and thereby learn something about each other. The next step is to request the participants to retell each other's stories using first-person voice ('I') and beginning with the statement, 'This is what I heard' (Pollock, 2006, p. 90). The assumption is that the retelling of the story facilitates the incorporation of others' memories into our own self and ours into others' through re-performance (Adams et al., 2015; Pollock, 2006). Unlike oral history, we, social workers need not play the role of the interviewer in storytelling settings. Instead, we have to play a facilitative role in the debriefing sessions that follows the storytelling session.

While working with local communities, diverse participatory action research projects have found meaning in exploring *cultural artefacts* such as local texts, wall journals, paintings, songs, theatre, artisanal farming and fishing tools and rituals associated with the use of these artefacts. Clarke (2005) observes that these cultural artefacts as implicated actants can be physically and/or discursively present in the situation of inquiry. These artefacts are routinely and discursively constructed by human actors from their own perspectives. Clarke (2005, p. 47) asks us to analyse 'Who is discursively constructing what, and how and why are they doing so?' Such an inquiry also helps us to understand the location and dynamics of power emerging in the contexts of implementing pilot projects. We could encourage community actors to set up local museums showcasing their past and present lives in the context of climate change adaptation. Cultural artefacts and other arts-based methods can aid better appreciation of pilots as they capture the reflexive, insightful and creative capacities of participants (Wilson and Flicker, 2014). These methods can be analysed by themselves or in congruence with other forms of data sets, as per the needs of the pilot project (ibid). Similarly, painting, drawing, mural making, drama and performance, collage, poetry, other forms of creative writing, puppetry and music creation are some of the popular arts-based strategies that can be applied in the contexts of adaptive innovation. The personal reflections of participants are embedded in the artefacts through a process of imaginative shaping, which also acts as a means of sharing insights on the pilot project with other actors (George, 2014). On many occasions, these artefacts also represent one of the significant action research outcomes. Cultural artefacts such as paintings, murals or poetry aim at sharing certain specific experiences as a narrative and are developed by participants to highlight a particular aspect or event, offering a powerful means of acknowledging and remembering them (Reece, 2014).

In a similar vein, *photography* is an extremely versatile method that community actors can use in a variety of ways to express their feelings, images or hopes pertaining to the pilots or future scenarios (Keegan, 2008). In recent years, *digital storytelling* has evolved as an audio-visual method for collecting participant-generated data on human experience. Digital storytelling adds the layer of meaning so that photographs and visuals become ways of living and telling one's stories of experience (Bach, 2008). The participants can be loaned small, easy-to-use camcorders and asked to make video diaries of their everyday lives pertaining to the pilot project. Through digital storytelling, participants produce raw, naive visual narratives of variable production quality but contain first-hand renderings of human experience, perceptions and behaviours (Rich, 2008). The primary mandate of participant actors is to tell their life stories in their own ways, showing and telling the aspects that best reveal their experiences. We could ask participants to record tours of their project sites and surroundings, their day-to-day activities sustaining adaptation and how they are adhering to or varying from the working models. To reveal the reflective and subjective nature of their piloting experience, participants are encouraged to speak directly to their camcorders each day as if they were writing in diaries, relating their experiences and responses, thoughts and feelings about those events.

Phase 6: Emergence

Emergence could create newer pathways of innovation in adaptation. Community actors should be able to develop their own capacities to solve complex adaptive challenges. However, community-based capacity building has to be understood as a continuous, non-linear, emergent and complex change process (Gilpil-Jackson, 2015). Facilitating a dialogue between diverse community actors and enabling them to manage trade-offs, find alternatives and at the same time ensure ethical adaptation measures become crucial in this context. Towards this end, we have to create spaces to host shared conversations to deliberate upon and understand the emergence of new, better relationships and patterns of organisation. Some of the methods that would help us to locate emergence are narrative interviews and life stories.

A *narrative interview* is an interview that is organised to capture actors' lived experiences and further interpreted through a narrative analysis. Narrative interviews are often organised temporally in the manner of life stories. Community actors often relate experiences as stories or narratives. While sharing their experiences, they tend to select and order events in ways that both reflect their own meanings and convey those meanings to others. Interviewing through the exploration of life stories can contribute meaningful insights to the adaptive innovation phases. Life stories have the potential to reveal the subjective reality of actors as well as the inner life of the interviewee, his or her moral struggles in being situated in a complex world, and previous successes and failures in adaptation (Faraday and Plummer, 1979; Fielding,

2003). They also help us to understand how actors dealt with moments of indecisions, confusions and ironies in their everyday experiences of adapting to climate change (ibid). These methods are useful for participant actors in reflecting upon the historical processes, ambiguities and structural constraints associated with climate change adaptation from the lens of their own or their peers' lived experiences.

The use of new media technologies to blend oral narratives in visual platforms to tell and share stories in a participatory manner could also help us observe the emerging patterns (Flicker and Hill, 2014). A *digital story* is a short film that is created through a combination of voice recording, still images, video clips, music or audio and text (ibid). At the emergence phase of adaptive innovation, these stories will be largely autobiographical in nature and would focus on the lived experiences of community actors. These digital stories could be made, edited and shared by community actors themselves. The emphasis here is that these actors themselves take decisions about what to say or show. The participatory process of creating stories is thus an important element of digital storytelling (Bach, 2008). These stories could be created and edited in workshop formats spread across a period of time. The shared photographs and conversations are viewed and re-viewed over time where the participant actors reflect, meditate and inquire into the emerging patterns of adaptation and self-organisation.

Ethnodrama can also be used as an innovative technique to study emergence. An ethnodrama or performed ethnography is the written transformation and adaptation of ethnographic research data into a dramatic play-script. These play-scripts are read aloud by a group of participants or performed for community actors as a live, public theatrical performance (Goldstein, 2014). The dramatic script consists of narratives that were collected through interviews, observation field notes, journal entries, diaries, media articles and so on (ibid). The actors and audiences for ethnodrama work are usually the community actors, the narratives of whose lived experiences form the core of the script. In the ethnodrama, both the performers and audience co-create the act (Saldaña, 2008). Some ethnodrama may be scripted as verbatim, slice-of-life naturalism to replicate authentic social interaction onstage. Others may be scripted and produced as direct address presentations, incorporating theatrical devices such as abstract movement, poetic choral speech, projected media and evocative background music (ibid).

The diverse methods of inquiry and practice that could be applied during each phase of the adaptive innovation project can be pictorially represented as shown in Figure 6.1.

Lines of inquiry

Some of the key lines of inquiry that could guide our practice are listed below. The guiding framework for these questions has to be that of ethical adaptation that focuses on the structural and relational factors of social inequalities, social exclusion

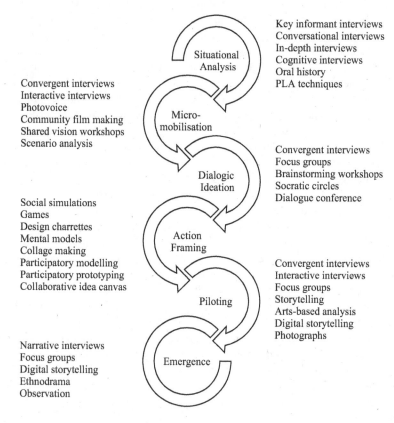

Key informant interviews
Conversational interviews
In-depth interviews
Cognitive interviews
Oral history
PLA techniques

Convergent interviews
Interactive interviews
Photovoice
Community film making
Shared vision workshops
Scenario analysis

Situational
Analysis

Micro-
mobilisation

Convergent interviews
Focus groups
Brainstorming workshops
Socratic circles
Dialogue conference

Dialogic
Ideation

Social simulations
Games
Design charrettes
Mental models
Collage making
Participatory modelling
Participatory prototyping
Collaborative idea canvas

Action
Framing

Convergent interviews
Interactive interviews
Focus groups
Storytelling
Arts-based analysis
Digital storytelling
Photographs

Piloting

Narrative interviews
Focus groups
Digital storytelling
Ethnodrama
Observation

Emergence

Figure 6.1 Methods of inquiry and practice in adaptive innovation.

and differential vulnerabilities, which shape adaptive capacities and barriers for specific individuals and groups of people in specific ways (Shackleton et al., 2015).

Situational Analysis: During this phase, we have to engage with diverse groups of community actors to reflect upon certain actionable questions such as:

- What are the prevalent adaptation strategies by the specific group of community actors?
- Why are these groups of community actors following this particular adaptation strategy rather than many other options?
- Are these strategies path dependent and have a historical continuity of rigid social structures and belief systems?
- What are the drivers that can enable social innovation and just adaptation in their specific situated contexts?
- What are the barriers that can act as impediments to innovation and adaptation and how can they be overcome?

- In this regard, what are the root causes of these barriers and what effects do they have on present and future pathways of adaptation?
- How can these cause and effects be transformed into achievable objectives?

Micro-mobilisation: Once the community actors are organised as a group or a collective entity, we should gradually work with them to develop a shared vision. Drawing insights from Southern (2015), some of the lines of inquiry that could facilitate a shared visioning process can be listed as follows:

- What stories can be narrated to describe the current situation?
- How are they relevant in the context of climate change adaptation?
- What success stories do they have to share?
- Do they generate excitement to develop future pathways?
- What are the stated values of the collective and how are they shared across the group?
- What will be the long-term aim of the collective?
- Do all community actors who are part of the collective share the same vision?
- Is there a synergy between the personal goals and the collective's goals?
- What are the unique strengths and limitations of the collective?
- Do they generate excitement or constrain people's willingness to develop future pathways?
- What is our role in the change process? Are they in synergy to the goals of the collective?

Dialogic Ideation: Some lines of inquiry that could facilitate dialogic ideation among community actors are:

- What situations require immediate attention and focus?
- How can we make sense of these situations and shape future pathways?
- What is specifically happening in the social-ecological system that is prompting the participant actors to explore alternatives?
- How are the local knowledge systems and practices of each actor going to shape the collective's imaginations?
- What structural and institutional processes reinforce these patterns of behaviour?
- What opportunities exist to imagine new ways of organising and interrelating?
- What are strengths of community actors to pursue these new ways of self-organising and strategic action?
- How are the participant actors reflecting upon their capacity to deal with the complexities in the social-ecological system?
- What additional capacities do we require for the same?
- What kind of narratives do community actors engage with when thinking and deliberating upon alternatives?
- What are the underlying beliefs and assumptions that are present in these narratives?

- Have we missed out anything crucial from these narratives and conversations so far?
- What is that we are not seeing and why?
- How can we co-create narratives that would display and communicate a shared imagination of the future?

As the inquiry progresses further, we should also explore the following aspects:

- What is the nature of actionable ideas that are floated upon?
- Can the action based on the idea be undertaken immediately?
- What value systems drive these processes of ideation?
- Whose interests and knowledge do these ideas represent?
- Do they recognise the voices of marginalised and subjugated actors?
- What are the gaps between espoused values and actionable ideas? Why do these gaps arise?
- How can these gaps be addressed?
- What capacities and resources do we have to move towards implementing the selected ideas as a working model or a pilot project?
- Does it require resources? Can we find the necessary resources ourselves? If yes, how?
- What are the enabling factors that are contributing to shared imaginations and conversational flows among diverse community actors?
- What are the constraining factors that are preventing community actors to involve in shared imagination and conversational flows?
- What kind of a dialogical process is further required to reflect and act up on the constraining and enabling factors?
- What are the contexts of co-creation and co-learning that are evolving as community actors come together for implementing their ideas into action?
- What role does the social worker envisage for himself/herself to sustain the processes of dialogic ideation and reflective practice?
- What strategies need to be developed to align diverse community actors in ways that sustain meaningful dialogue and leverage for change?
- What scenarios do we need to anticipate as we plan for change?

Action Framing: The following lines of inquiry will be helpful to facilitate action framing in a reflective and co-creative manner. In this regard, the larger frame of thought can be borrowed from the questions that Haraway had raised in her work on situated knowledge,

> How to see, where to see from? What limits the vision? What to see for? Whom to see with? Who gets to have more than one point of view? Who gets blinded? Who wears blinders? Who interprets the field? What other sensory powers do we want to cultivate besides the vision?
>
> (Haraway, 1988, p. 587)

Other specific lines of inquiry include:

- What would it look like if this idea were a reality?
- What knowledge and skills do community actors already have to develop experiential or visual working models?
- What knowledge, skills and abilities do they need to build?
- What are the questions and assumptions that need to be answered or verified?
- What are we really trying to do?
- What resources are required to develop and implement a particular working model?
- What innovative ways already exist that would make this working model as a possible model to be replicated?
- Are community actors in similar social-ecological systems already implementing the suggested adaptation strategies or model?
- Why did similar models fail in the past? Or why did they succeed?
- What do you want to learn through the action framing process? Or what is the significant learning from the process?
- Has the process motivated and inspired community actors to sustain the adaptive innovation process?
- What is the kind of feedback that community actors in specific social worlds are signalling to one another?
- Will the actionable project address the needs of subjugated and marginalised actors? Who else will be benefited with such a model?
- Will there be anyone at a disadvantage if the model is implemented?
- What improvisations are further required to strengthen the adaptive capacities?
- What aspects need further thinking, refining and exploration?

Piloting: Some of the key lines of inquiry that could facilitate the reflective analysis of pilot projects are as follows:

- What are the indicators of success and failure of the pilot project?
- How can we understand them?
- How should we document the learning process?
- Do we have some baseline information to compare our present understanding?
- Who are the diverse community actors defining the project as a success or a failure?
- What are their values, interests, knowledge and power relations that shape the understanding of the project as a success or a failure?
- What adjustments should we carry out in response to the insights that community actors have gathered from the pilot?
- Does the pilot have the potential to scale?
- Should we scale it?
- What will be the implications of the scaling up process in terms of the vision, costs, actions and outcomes?

- What can be other unintended consequences of scaling up?
- What kind of institutional interface should we build for successful scaling up?
- What kind of adjustments should we make in the present model to scale it up?
- How are we looking out and within for the feedback?

Emergence: Our pursuit to understand the nature of emergence could be supported through certain lines of inquiry as given below.

- How are different participant actors making sense of the situation? And what makes sense?
- How are different working models and pilots unfolding at the grassroots level and how are they getting adapted in the social-ecological system?
- What are the intended and unintended consequences of our intervention?
- Are all participant actors still committed to the values of justice, care and solidarity?
- How are power relations unfolding in the emerging situation?
- What is the nature of self-organisation that is emerging among diverse actors involved in adaptive innovation?
- What is the nature of emerging social networks, communication and relationships of trust and leadership among participant actors?
- What has disappeared when new knowledge, practice and institutions have emerged?
- Are participant actors owning up the consequences – both successes and failures?
- What are the decisions of participant actors regarding future cycles of adaptive innovation? What are the consequences of these decisions on the originally stated adaptation design?
- What is the nature of diffusion of innovations, and how are they being replicated?
- What have we learned from the whole practice context?
- How can we improve our practice?

It is important to remember that all the methods discussed in this chapter do not stand alone or in isolation with other methods during the adaptive innovation process. These methods can overlap with one another. Each method can be modified and applied suitably in any of the other phases as well. The guidelines and lines of inquiry discussed under each method or technique are neither rigid nor fixed. They can be altered or completely avoided based on the situated knowledge and interests of community actors. What matters most importantly while applying these methods of inquiry and practice is that they should aid in reflective practice and facilitate people-centred decision-making in climate change adaptation.

References

Adams, T.E., Jones, S.H., and Ellis, C. (2015). *Autoethnography*, New York: Oxford University Press.

Alderfer, C.P., and Smith, K.K. (1982). Studying intergroup relations embedded in organisations, *Administrative Science Quarterly*, 27(1), pp. 35–65.

Angrosino, M.V. (2004). Disclosure and interaction in a monastery, in L. Hume and J. Mulcock (Eds.). *Anthropologists in the field: cases in participant observation*, New York: Columbia University Press, pp. 18–31.

Arendt, H. (1958). *The human condition*, Chicago, IL: University of Chicago Press.

Awuor, C., and Hammill, A. (2009). Rain calendars: a tool for understanding changing rainfall patterns and effects on livelihoods, in H. Reid, M. Alam, R. Berger, T. Cannon, S. Huq, and A. Milligan (Eds.). *Community-based adaptation to climate change, participatory learning and action 60*, London: IIED, pp. 149–153.

Bach, H. (2008). Visual narrative inquiry, in L.M. Given (Ed.). *The Sage encyclopedia of qualitative research methods*, Vol. 1&2, New Delhi: Sage, pp. 938–940.

Barkved, L., de Bruin, K., and Romstad, B. (2014). *Mapping of drought vulnerability and risk*, Final report on WP 2.3: Extreme Risks, Vulnerabilities and Community based-Adaptation in India (EVA): A Pilot Study, New Delhi: CIENS-TERI, TERI Press.

Barndt, D. (2014). Photovoice, in D. Coghlan and M. Brydon-Miller (Eds.). *The Sage encyclopedia of action research*, New Delhi: Sage, pp. 620–623.

Bizikova, L., Dickinson, T., and Pintér, L. (2009). Participatory scenario development for climate change adaptation, in H. Reid, M. Alam, R. Berger, T. Cannon, S. Huq, and A. Milligan (Eds.). *Community-based adaptation to climate change, participatory learning and action 60*, London: IIED, pp. 167–172.

BUIOH. (2016). Introduction to oral history, Baylor University Institute for Oral History. Retrieved from www.baylor.edu/oralhistory [Last accessed on 11 September 2019].

Busse, M. (2016). Socratic circles: enhanced creativity and ideation through participant-centred dialogue, 16 December 2016, HCMA. Retrieved from https://hcma.ca/socratic-circles/ [Last accessed on 11 September 2019].

Caulkins, D.D. (2014). Ethnography, in D. Coghlan and M. Brydon-Miller (Eds.). *The Sage encyclopedia of action research*, New Delhi: Sage, pp. 309–314.

Chaitin, J. (2008). Oral history, in L.M. Given (Ed.). *The Sage encyclopedia of qualitative research methods*, Vol. 1&2, New Delhi: Sage, pp. 583–585.

Chambers, R. (1994). The origins and practices of participatory rural appraisal, *World Development*, 22(7), pp. 953–969.

Chambers, R. (1997). *Whose reality counts? Putting the first last*, London: Intermediate Technology Publications.

Chevalier, J.M., and Buckles, D.J. (2008). *SAS2 a guide to collaborative inquiry and social engagement*, New Delhi: Sage.

Clandinin, D.J., and Caine, V. (2008). Narrative inquiry, in L.M. Given (Ed.). *The Sage encyclopedia of qualitative research methods*, Vol. 1&2, New Delhi: Sage, pp. 541–544.

Clarke, A.E. (2005). *Situational analysis: grounded theory after the postmodern turn*, New Delhi: Sage.

Cunningham, J. (2014). Metaphor, in D. Coghlan and M. Brydon-Miller (Eds.). *The Sage encyclopedia of action research*, New Delhi: Sage, pp. 534–536.

Di Gessa, S. (2008). *Participatory mapping as a tool for empowerment: experiences and lessons learned from the ILC network*, Rome, Italy: International Land Coalition.

Dick, B. (1990). *Convergent interviewing*, Brisbane: Interchange.

Driedger, S.M. (2008). Convergent interviewing, in L.M. Given (Ed.). *The Sage encyclopedia of qualitative research methods*, Vol. 1&2, New Delhi: Sage, pp. 125–127.

Driedger, S.M. (2014). Convergent interviewing, in D. Coghlan and M. Brydon-Miller (Eds.). *The Sage encyclopedia of action research*, New Delhi: Sage, pp. 186–187.

Dunn, K. (2017). Prototyping models of climate change: new approaches to modelling climate change data, *Doctoral Dissertation*, Faculty of Architecture, Design and Planning: The University of Sydney.

Egmose, J. (2015). *Action research for sustainability: social imagination between citizens and scientists*, London: Routledge.

Ellen, G.J., Leeuwen, C., Kuindersma, W., Breman, B., and Lamoen, F. (2015). Adaptive governance in practice: a learning approach based on action research designed for the implementation of climate adaptation measures, in A. Buuren, J. Eshuis, and M. Vilet (Eds.). *Action research for climate change adaptation: developing and applying knowledge for governance*, London: Routledge, pp. 112–129.

Ellis, C.S. (2008). Interacting interview, in L.M. Given (Ed.). *The Sage encyclopedia of qualitative research methods*, Vol. 1&2, New Delhi: Sage, pp. 443–445.

Elsawah, S., Guillaume, J.H.A., Filatova, T., Rook, J., and Jakeman, A.J. (2015). A methodology for eliciting, representing, and analysing stakeholder knowledge for decision making on complex socio-ecological systems: from cognitive maps to agent-based models, *Journal of Environment Management*, 151, pp. 500–516.

Faraday, A., and Plummer, K. (1979). Doing life histories, *Sociological Review*, 27(4), pp. 773–798.

Fetterman, D.M. (2008a). Key informant, in L.M. Given (Ed.). *The Sage encyclopedia of qualitative research methods*, Vol. 1&2, New Delhi: Sage, p. 477.

Fetterman, D.M. (2008b). Ethnography, in L.M. Given (Ed.). *The Sage encyclopedia of qualitative research methods*, Vol. 1&2, New Delhi: Sage, pp. 288–292.

Fielding, N. (Ed.). (2003). *Interviewing*, Vol. II, New Delhi: Sage.

Flicker, S. (2014a). Collaborative data analysis, in D. Coghlan and M. Brydon-Miller (Eds.). *The Sage encyclopedia of action research*, New Delhi: Sage, pp. 121–124.

Flicker, S. (2014b). Disseminating action research, in D. Coghlan and M. Brydon-Miller (Eds.). *The Sage encyclopedia of action research*, New Delhi: Sage, pp. 276–280.

Flicker, S., and Hill, A. (2014). Digital storytelling, in D. Coghlan and M. Brydon-Miller (Eds.). *The Sage encyclopedia of action research*, New Delhi: Sage, pp. 266–270.

George, A. (2014). Aesthetics, in D. Coghlan and M. Brydon-Miller (Eds.). *The Sage encyclopedia of action research*, New Delhi: Sage, pp. 29–31.

Gieselman, R.E., Fisher, R.P., Mackinnon, D.P., and Holland, H.L. (1986). Enhancement of eyewitness memory with the cognitive interview, *American Journal of Psychology*, 99(3), pp. 385–401.

Gilpil-Jackson, Y. (2015). Transformative learning during dialogic organisation development, in G.R. Bushe and R.J. Marshak (Eds.). *Dialogic organisation development: the theory and practice of transformational change*, Oakland, CA: Berrett-Koehler Publishers, pp. 245–267.

Goldstein, T. (2014). Performed ethnography, in D. Coghlan and M. Brydon-Miller (Eds.). *The Sage encyclopedia of action research*, New Delhi: Sage, pp. 612–613.

Greenwood, D.J., and Levin, M. (2007). *Introduction to action research: social research for social change*, New Delhi: Sage.

Haraway, D. (1988). Situated knowledges: the science question in feminism and the privilege of partial perspective, *Feminist Studies*, 14(3), pp. 575–599.

Hollway, W., and Jefferson, J. (1997). Eliciting narrative through in-depth interview, *Qualitative Inquiry*, 3(1), pp. 53–70.

Hughes, R. (2008). Vignettes, in L.M. Given (Ed.). *The Sage encyclopedia of qualitative research methods*, Vol. 1 & 2, New Delhi: Sage, pp. 918–920.

Huntjens, P., Ottow, B., and Lasagne, R. (2015). Participation in climate adaptation in the Lower Vam Co river basin in Vietnam, in A. Buuren, I. Eshuis, and M. Vilet (Eds.). *Action research for climate change adaptation: developing and applying knowledge for governance*, London: Routledge, pp. 55–75.

Jackson, C. (2015). Facilitating collaborative problem solving with human-centred design: the making all voices count governance programme in 12 countries of Africa and Asia, *Knowledge Management for Development Journal*, 11(1), pp. 91–106.

Johnson, K., and Porter, S. (2014). Disabled people's organisation, in D. Coghlan and M. Brydon-Miller (Eds.). *The Sage encyclopedia of action research*, New Delhi: Sage, pp. 270–273.

Kasper, A.S. (1994). A feminist, qualitative methodology: a study of women with breast cancer, *Qualitative Sociology*, 17(3), pp. 263–281.

Keegan, S. (2008). Photographs in qualitative research, in L.M. Given (Ed.). *The Sage encyclopedia of qualitative research methods*, Vol. 1 & 2, New Delhi: Sage, pp. 619–622.

Kitzinger, J. (1994). The methodology of focus groups: the importance of interaction between research participants, *Sociology of Health and Illness*, 16(1), pp. 103–121.

Ledwith, M. (2017). Emancipatory action research as a critical living praxis: from dominant narratives to counternarratives, in L.L. Rowell, C.D. Bruce, J.M. Shosh, and M.M. Riel (Eds.). *The Palgrave international handbook of action research*, New York: Palgrave Macmillan, pp. 49–62.

Liamputtong, P. (2014a). Oral history, in D. Coghlan and M. Brydon-Miller (Eds.). *The Sage encyclopedia of action research*, New Delhi: Sage, pp. 574–575.

Liamputtong, P. (2014b). Experiential knowing, in D. Coghlan and M. Brydon-Miller (Eds.). *The Sage encyclopedia of action research*, New Delhi: Sage, pp. 323–325.

Logie, C. (2014). Focus groups, in D. Coghlan and M. Brydon-Miller (Eds.). *The Sage encyclopedia of action research*, New Delhi: Sage, pp. 355–358.

Mack, N., Woodsong, C., MacQueen, K.M., Guest, G., and Namey, E. (2005). *Qualitative research methods: a data collector's field guide*, Durham, NC: Family Health International.

Martínez-Harms, M.J., and Balvanera, P. (2012). Methods for mapping ecosystem service supply: a review, *International Journal of Biodiversity Science, Ecosystem Services & Management*, 8(1–2), pp. 17–25.

Matthews, J.R. (2017). Understanding indigenous innovation in rural West Africa: challenges to diffusion of innovations theory and current social innovation practice, *Journal of Human Development and Capabilities*, 18(2), pp. 223–238.

McKernan, J. (1991). *Curriculum action research: a handbook of methods and resources for the reflective practitioner*, London: Routledge Falmer.

McNiff, J., Lomax, P., and Whitehead, J. (1996). *You and your action research project*, London: Hyde Publications.

Mead, G. (2014). Storytelling, in D. Coghlan and M. Brydon-Miller (Eds.). *The Sage encyclopedia of action research*, New Delhi: Sage, pp. 728–731.

Morgan, D.L. (1996). Focus groups, *Annual Review of Sociology*, 22, pp. 129–152.

Narayanasamy, N. (2009). *Participatory rural appraisal: principles, methods and applications*, New Delhi: Sage.

Ngwenya, H., and Hagmann, J. (2011). Making innovation systems work in practice: experiences in integrating innovation, social learning and knowledge in innovation platforms, *Knowledge Management for Development Journal*, 7(1), pp. 109–124.

Norris, J. (2008). Collage, in L.M. Given (Ed.). *The Sage encyclopedia of qualitative research methods*, Vol. 1&2, New Delhi: Sage, pp. 94–97.

Noyes, E. (2018). Teaching entrepreneurial action through prototyping: the prototype-it challenge, *Entrepreneurship Education and Pedagogy*, 1(1), pp. 118–134.

Osterwalder, A., and Pigner, Y. (2009). *Business model generation*, Self-published. ISBN: 978-2-8399-0580-0.

Piccolella, A. (2013). *Adaptation in practice: increasing adaptive capacity through participatory mapping*, Rome, Italy: Environment and Climate Division: IFAD.

Pollock, D. (2006). Memory, remembering, and histories of change, in S. Madison and J. Hamera (Eds.). *The Sage handbook of performance studies*, Thousand Oaks, CA: Sage, pp. 87–105.

Poole, J.M., and Mauthner, O. (2014). Interviews, in D. Coghlan and M. Brydon-Miller (Eds.). *The Sage encyclopedia of action research*, New Delhi: Sage, pp. 463–465.

Pound, B., and Conroy, C. (2017). The innovation systems approach to agricultural research and development, in S. Snapp and B. Pound (Eds.). *Agricultural systems: agroecology and rural innovation for development*, Amsterdam, The Netherlands: Elsevier and Academic Press, pp. 371–405.

Reece, J. (2014). Narrative, in D. Coghlan and M. Brydon-Miller (Eds.). *The Sage encyclopedia of action research*, New Delhi: Sage, pp. 547–549.

Rich, M. (2008). Video intervention/prevention assessment, in L.M. Given (Ed.). *The Sage encyclopedia of qualitative research methods*, Vol. 1&2, New Delhi: Sage, pp. 914–916.

Riessman, C.K. (2008). Narrative analysis, in L.M. Given (Ed.). *The Sage encyclopedia of qualitative research methods*, Vol. 1&2, New Delhi: Sage, pp. 539–540.

Ritchie, D.A. (2003). *Doing oral history: a practical guide*, New York: Oxford University Press.

Ritchie, D.A. (2015). *Doing oral history*, 3rd edition, New Delhi: Oxford.

Roehrig, M.J., Schwendenwein, J., and Bushe, G.R. (2015). Amplifying change: a three-phase approach to model, nurture, and embed ideas for change, in G.R. Bushe and R.J. Marshak (Eds.). *Dialogic organisation development: the theory and practice of transformational change*, Oakland, CA: Berrett-Koehler Publishers, pp. 325–348.

Roggema, R., Martin, J., and Vos, L. (2015). Governance of climate adaptation in Australia: design charrettes as a creative tool for participatory action research, in A. Buuren, J. Eshuis, and M. Vilet (Eds.). *Action research for climate change adaptation: developing and applying knowledge for governance*, London: Routledge, pp. 94–111.

Roulston, K.J. (2008). Conversational interviewing, in L.M. Given (Ed.). *The Sage encyclopedia of qualitative research methods*, Vol. 1&2, New Delhi: Sage, pp. 127–129.

Rouse, W.B., and Morris, N.M. (1986). On looking into the black box: prospects and limits in the search for mental models, *Psychological Bulletin*, 100(3), pp. 349–363.

Saldaña, J. (2008). Ethnodrama, in L.M. Given (Ed.). *The Sage encyclopedia of qualitative research methods*, Vol. 1&2, New Delhi: Sage, pp. 283–285.

Sanders, E.B.N. (2013). Prototyping for the design spaces of the future, in L. Valentine (Ed.). *Prototype: design and craft in the 21st century*, London: Bloomsbury, pp. 59–74.

Schenk, T., and Susskind, L. (2015). Using role-play simulations to encourage adaptation: serious games as tools for action research, in A. Buuren, J. Eshuis, and M. Vilet (Eds.). *Action research for climate change adaptation: developing and applying knowledge for governance*, London: Routledge, pp. 148–163.

Senge, P.M. (1990). *The fifth discipline: the art and science of the learning organization*, New York: Currency Doubleday.

Senge, P.M., Kleiner, A., Roberts, C., Ross, R.B., and Smith, B.J. (2010). *The fifth discipline fieldbook: strategies and tools for building a learning organisation*, London: Nicholas Brealey Publishing.

Shackleton, S., Ziervogel, G., Sallu, S.M., Gill, T., and Tschakert, P. (2015). Why is socially just climate change adaptation in sub-Saharan Africa so challenging? A review of barriers identified from empirical cases, *Wiley Interdisciplinary Reviews: Climate Change*, 6(3), pp. 321–344.

Shotter, J. (2014). Dialogue conferences, in D. Coghlan and M. Brydon-Miller (Eds.). *The Sage encyclopedia of action research*, New Delhi: Sage, pp. 260–265.

Sommer, B.W., and Quinlan, M.K. (2009). *The oral history manual*, Lanham, MD: Altamira Press.

Southern, N. (2015). Framing inquiry: the art of engaging great questions, in G.R. Bushe and R.J. Marshak (Eds.). *Dialogic organisation development: the theory and practice of transformational change*, Oakland, CA: Berrett-Koehler Publishers, pp. 269–289.

Springett, J. (2014). Interactive research, in D. Coghlan and M. Brydon-Miller (Eds.). *The Sage encyclopedia of action research*, New Delhi: Sage, pp. 451–452.

Stringer, E.T. (2007). *Action research*, New Delhi: Sage.

Swaans, K., Boogaard, B., Bendapudi, R., Taye, H., Hendrickx, S., and Klerkx, L. (2014). Operationalizing inclusive innovation: lessons from innovation platforms in livestock value chains in India and Mozambique, *Innovation and Development*, 4(2), pp. 239–257.

Swart, C. (2015). Coaching from a dialogic organisational development paradigm, in G.R. Bushe and R.J. Marshak (Eds.). *Dialogic organisation development: the theory and practice of transformational change*, Oakland, CA: Berrett-Koehler Publishers, pp. 349–370.

Tassinari, V., Piredda, F., and Bertolotti, E. (2017). Storytelling in design for social innovation and politics: a reading through the lenses of Hannah Arendt, *The Design Journal*, 20(sup1), pp. S3486–S3495.

Thompson, P. (1978). *The voice of the past oral history*, Oxford: Oxford University Press.

Tschakert, P., and Sagoe, R. (2009). Mental models: understanding the causes and consequences of climate change, in H. Reid, M. Alam, R. Berger, T. Cannon, S. Huq, and A. Milligan (Eds.). *Community-based adaptation to climate change, participatory learning and action 60*, London: IIED, pp. 154–159.

Veale, S., and Schilling, K. (2004). *Talking history: oral history guidelines*, Hurstville: Department of Environment and Conservation, National Library of Australia.

Westerlund, B., and Wetter-Edman, K. (2017). Dealing with wicked problems, in messy contexts, through prototyping, *The Design Journal*, 20(sup1), pp. S886–S899.

WHO and ExpandNet. (2011). *Beginning with the end in mind: planning pilot projects and other programmatic research for successful scaling up*, Geneva: WHO and ExpandNet.

Wilson, C., and Flicker, S. (2014). Arts-based action research, in D. Coghlan and M. Brydon-Miller (Eds.). *The Sage encyclopedia of action research*, New Delhi: Sage, pp. 58–62.

Winter, R., and Munn-Giddings, C. (2001). *A handbook for action research in health and social care*, London: Routledge.

Zandee, D.P. (2014). Appreciative inquiry and research methodology, in D. Coghlan and M. Brydon-Miller (Eds.). *The Sage encyclopedia of action research*, New Delhi: Sage, pp. 48–50.

7 Actor interface analysis and reflective practice

Introduction

The previous chapter had elaborated the diverse methods of inquiry and practice involved in adaptive innovation. The use of these diverse methods also highlights the need to have a suitable analytical frame that will help us to reflect on our own and other actor's knowledge-action frames. The analytical framework should be capable of helping us to reflect on our own subjectivity, to trust feelings and contain them suiting the contexts of respective social encounters (Papell, 1996). A way forward will be to recognise the nature and significance of diverse actor interfaces in climate change adaptation. Social interface analysis represents the dynamic and conflictive nature of social encounters between diverse social actors with differing interests, knowledge, resources and power in various social realms (Long, 1989; Long and Long, 1992). Such an actor-oriented framework acts as a reflective engagement that would enable us to facilitate each phase of adaptive innovation based on the differentiated meanings and strategies that diverse actors evolve during the process. This chapter discusses this analytical framework in detail.

Analysing social interfaces in adaptive innovation

Social encounters in the context of this book refer to the spatial and networked flows of interactions among social actors at each phase of the adaptive innovation process. We should look out specifically for those social encounters that seek to actively connect the ideas, imaginations and actions of community actors towards ethical adaptation. The social encounters signify the dynamic nature of adaptive innovation, which is shaped and re-shaped by the discursive practices of community actors. These actors are often situated within the boundaries of their social worlds (manifested through the expressions of their values, interests and knowledge) and constrained or enabled by the nature of relationships within it (manifested in the form of embedded power relations). Climate change adaptation would often require the involvement of multiple networks of actors engaged in different types of social encounters. In each of these social encounters, we could locate the differentiated and networked nature of knowledge and practice. Certain social encounters would depict that the change itself is slow and will be resisted by community actors. Other sets of social encounters could indicate that the change will be fast, drastic, emergent or radical. Each social encounter is thus a dynamic space, which is altered by the community actors themselves, individually or collectively through their spontaneous or pre-meditated actions. We have to reflect on specific social encounters and discern the underlying patterns, similarities and variations, reflect and act accordingly. It shows the paths towards convergent actions, which are also spaces where sense of owning up and sharing responsibilities emerges at a larger collective level (Clarke, 2005). An understanding of how these social encounters and actor networks are recursively constructed over a period of time will add value to the adaptive innovation process.

The actor interface analysis would enable us to analyse those social encounters that were strategic in terms of how different actors agreed upon, negotiated or contested while arriving at decisions pertaining to adaptation. It also helps us to interrogate the diverse situations that led to the emergence of novel objectives, innovative relationships, actionable knowledge and related action. Equally important is our ability to understand the emergence of different types of power relations during the adaptation process. We have to be alert to that nature of embedded power that intersects with aspects of gender, caste, class, ethnicity, sexuality, age, ability and many other structural factors that produce and sustain inequalities among actors in different adaptation situations. We also need to be aware of the fact that the participant actors involved in adaptive innovation processes have to consistently engage and negotiate with plural power structures (Viterna, 2013). Power refers to the capacity of an actor or actors to make a difference in some state of affairs or influence a particular situation, behaviour or mode of action. It indicates the transformative capacity of human action (ibid). The social meanings embedded in every social encounter are thus shaped by the values, interest, knowledge and power of different intersectional actors interacting at diverse institutional levels (refer to Figure 7.1). These institutional levels can be formal and informal in nature, represented such as the binaries of state versus community, scientific versus local, private versus commons and so on; or else these boundaries can be extremely blurred and ambiguous, representing the nature of hybrid or mixed institutions or even institutional bricolage (Cleaver, 2001).

Analysing power involves critically reflecting on the voices of diverse actors. While voices of privileged actors are often paid attention to, those of subjugated actors remain ignored (Clarke, 2005). In this regard, it is very important to emphasise on the analysis of differences among community actors across the layered and intersectional structures of power (ibid). The interface analysis therefore would critically examine the emergent bio-political implications of the adaptation process. One has to consistently reflect upon the historical power relations and their implications on present-day knowledge production, policymaking and practices pertaining to adaptation (Morchain, 2018). The analysis should give us clues on the governing ideologies, power

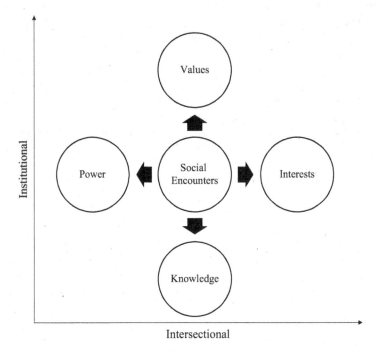

Figure 7.1 Actor interfaces in adaptation.

relations, interests and rationalities of diverse actors and practices in adaptation settings (Klepp and Chavez-Rodriguez, 2018). Such an analysis in itself could have multiple meanings and scopes due to the prevalent and emerging inequalities that arise in terms of access and implementation in diverse cultural contexts (Ulloa, 2018). Therefore, it is crucial not only to critically interrogate the discursive framings within which adaptation emerges, but also explore how people respond to the idea of adaptation itself (de Wit, 2018).

In this context, it is crucial to understand how diverse community actors take account of significant others and the nature of their affiliation to the innovation platforms. These observations will also help us to reflect on the extent to which community actors actually adhere to shared values and norms of their respective innovation platforms or examine the reasons why some actors are deviating from the expected norms of behaviour. Further, it is important to analyse how community actors manoeuvre these normative and institutional frameworks, so as to enforce their own values and pursue specific interests (Long and Long, 1992). The emphasis is towards reflecting up on the contexts that facilitate emergent action (Seur, 1992). For example, the complex negotiation processes involving cooperation, problem solving or conflict resolution among particular actors with specific interests could be understood in depth. It also helps us to see how our interventions are transformed at each phase of the adaptive innovation due to the negotiations between intervening parties. The social interface frame thus provides insights into the full situation of inquiry and practice.

In each phase of the adaptive innovation process, we have to be aware of how diverse values and beliefs, interests, knowledge and power of different actors' interface with each other. The response of community actors to each phase of adaptive innovation, and the situated nature of their strategic actions and consequences (in terms of both discontinuities and innovation) need to be given much attention. The key assumption guiding such an analysis is that external interventions including those of ours, as social workers, enter the lifeworlds of people whom we work with, and thus form part of those resources and constraints that they strategically develop (Long, 1992). These people, despite their contexts of vulnerability and poverty, engage in preserving some normative consensus and control over their own social arrangements in the face of both internal and external pressures (Arce and Long, 1987). In addition, they are also knowledgeable and capable of extracting some benefits from it (Scott, 1985). Such an analysis would also throw light on which actor networks matter and how they operate or manoeuvre in specific situations of practice (Long, 2001; Viterna, 2013). The endurance of the adaptive innovation processes itself could be understood through the social interface analysis, providing new answers to existing complex situations and how future innovations, coalescence of power and domination can be shaped or reshaped.

The needs, aspirations and capacities of relevant community actors have to be explored in depth. Diverse actors would maintain different strategies to meet their goals and justify their decisions. Moreover, these decision-making strategies are dynamic in nature, prompting actors to change their decisions and action strategies based on their own experiences and feedback from others. Thus, the collective outcomes of decisions made by community actors add up significantly to the complexities of the adaptation processes (Elsawah et al., 2015). We need to be alert and proactive to the shaping and reshaping of adaptation strategies and decisions at different phases of adaptive innovation. Some crucial lines of inquiry that could facilitate the analysis of actor interfaces are as follows:

- Who are the diverse actors involved in a particular social encounter? What are the characteristics of these social encounters? And how are they shaped by the dynamics of the specific social-ecological system?
- What are the diverse values and beliefs, interests, knowledge and power of different actors involved in specific social encounters? How have these dimensions shaped the capabilities of vulnerable groups?
- What are the consequences of actors' interfaces on people's assets, knowledge, skills and institutions in a specific social-ecological context?
- What has been the nature of response among community actors to the intervention of outside experts in the day-to-day activities of the innovation platform?

- What has been the nature of discontinuities and emergence associated with the innovation platform, specifically with reference to expert interventions from outside?
- How are frictions, disagreements and conflicts mediated and transformed at critical junctures of the social encounter?

An analysis of actor-oriented interactions and their interfaces provide ample opportunities to look at social work practice from the inside (Kirkwood et al., 2016). In each phase of adaptive innovation, we, as social workers will also be faced with inner struggles, as we encounter other actors with different and multiple identities, power, privilege or deprivation. Analysis of actor interfaces in adaptive innovation involves reflecting on our actions and linked knowledge outcomes in a systematic way. Further, the diverse themes that would emerge from an analysis of actor interfaces may help us to construct theories and methods emerging from our own contexts of practice. The insights gained would equip us to face new challenges and at the same time be open to broader innovative ideas that would feed into the adaptation process. Such an approach requires a systematic reflection of both process and outcomes, which is often bound by the action present. Such reflective practices have to occur in a time zone in which action can still make a difference to the situation (Schön, 1985; Somekh, 1995). Thus, it provides necessary value for us to adapt to the developing processes and continuously act in relation to the other actors in each social encounter. It also reveals the opportunities and constraints that would emerge in the context of multiple actors' situatedness and embedded power relations. This requires us to enter into a collaborative search for meaning with diverse community actors, listen to their voices, narratives and constructions of reality (Hartman, 1994).

S.F.F.7.1 Narratives on actor interfaces.

I would like to narrate here my observations of actor interfaces that have surfaced during different social encounters between traditional fishing communities and other stakeholders while designing adaptation strategies. Traditional fishing communities in India possess immense local knowledge in forecasting coastal hazards. At the same time, their situated understanding of coastal hazards is quite different from those of expert-driven scientific constructions. There are differences in the very language that is used to define and categorise coastal hazards. While external experts such as scientists and law enforcers use terms such as storms, cyclones, coastal erosion or tidal waves to define a particular coastal hazard, traditional fishing communities may have localised, but holistic conceptualisation that is very much embedded in their specific social-ecological system (Santha, 2015). And as we probe deeper, we realise that there are marked differences in the manner in which local communities and formal actors conceive and design risk reduction and related adaptation strategies. One could observe that these strategies are often founded on contrasting values, interests, knowledge and power among different actors. I have noticed that local fishing communities often consider their knowledge systems as efficient and reliable than formal knowledge systems. Once a group of traditional fishermen told me,

> We listen to warnings pertaining to change in weather, if it is issued by one among us. We trust the knowledge of our elders. They have more experience and can sense any change in the environment. We don't trust the warnings issued by the police, radio, television or the collector. They always prove to be false warnings and neither do they recognise our thinking and actions!

The expert-driven working models and technologies often do not blend with the lifeworld of traditional fishing communities. Often these dominating technologies are imposed by outsiders thinking that it will be good for the local communities. I happened to visit a fishing village, where the government had installed equipment that could warn coastal hazards such as sea surge and the tsunami. However, during my conversations with the local community actors, I realised that they were completely sceptical about such technologies. It seems that this particular equipment emitted false alarms and created panic in the community. In the words of a fisherwoman,

> That equipment often generates false alarm. Initially, we used to get scared. I used to run far away from the sea with my kids, livestock and poultry. But nothing happened. Then we realised that there is something wrong with this equipment. These days we neglect it!

Similar incidences have happened across different coastal villages. A fish-workers' trade union leader commented as follows:

> They (scientists and the Fisheries Department officials) have installed an early warning equipment in the office of our fish-workers' cooperative society. The officials never informed us that there is an alarm fixed in it. One night, the alarm went off. We were unable to access the equipment and switch off the alarm, as the concerned officials had locked the room and gone home for the day. We were awake the whole night wondering what the noise was! We were also informed that this equipment could help us in finding potential fishing grounds. We even ignore that information. Most of the display is in English that we do not understand.

As fisherfolk neglect such formal and techno-centric conceptualisations of risk, the state actors also avail certain tactical strategies to enforce these understandings on the former. Department officials, for instance, issue statements along with the warning that non-compliance could render fisherfolk ineligible for claiming any compensation to damage or loss due to coastal hazards.

Now, let us look into the nature of actor interfaces in riverine fisheries. Few state governments, scientific institutions and civil society organisations have also initiated resource enhancement programmes as an adaptation measure to boost fish production in the inland fisheries sector. Under this programme, fingerlings of certain exotic fish varieties are bred in captivity and released into the rivers. The state officials believe that this strategy could not only improve the livelihood security of fisherfolk, but also reduce the pressure on indigenous fish varieties. In contrast, the fisherfolk believe that the introduction of exotic fish varieties is harmful to their riverine ecosystem as they consider that in the long run, the exotic varieties could endanger the indigenous fish species. The fishermen believe that the state should encourage introduction of indigenous fish varieties to the respective rivers. According to them, these activities should be carried out at a specific time of the year, preferably during the natural spawning season. They also suggest that specific habitats should be identified to introduce these species, rather than randomly releasing them into the upper stretches of the river. In addition, fishing communities in the respective wetland regions observe that such processes lack transparency and community involvement. They also comment that there are possibilities of widespread corruption in the procurement and release of fingerlings. Countering these arguments, the state officials deny any form of corruption. Instead, they term the discrepancies in the quantity of fingerlings released into the rivers as 'sampling errors' during stocktaking.

Aquaculture is promoted as yet another adaptation strategy among farmers cultivating paddy in wetlands. However, traditional fisherfolk view the promotion of aquaculture in paddy fields with considerable apprehension. Various reasons can be attributed to this. The traditional inland fishermen fish from the rivers during the active season, and from the paddy fields during the non-agricultural season. Thus, the paddy fields are a kind of seasonal commons that ensures a means of livelihood security for traditional fisherfolk. The exclusive promotion of aquaculture in paddy fields has begun to erode the common pool nature of wetlands, and traditional fisherfolk are being prevented from fishing in these wetlands. The fisherfolk believe that aquaculture can accelerate the privatisation of commons, and at the same time prevent them from accessing their customary livelihood resources. In addition, fisherfolk believe that the present-day practices of aquaculture could erode the ecological security of wetlands. As more and more exotic and predator varieties of fish are introduced into the aquaculture ponds, these species will always remain as a threat to indigenous fish varieties living in the rivers and lakes. Any event of flooding or breach of ponds or tanks can disrupt the native biodiversity of the whole wetland ecosystem. The above narratives illustrate that adaptation designs have to be nurtured by institutional platforms that are glued to the fishing communities' social-ecological system. The strength of their local knowledge system is that these are routinised as part of their everyday livelihood struggles. Formal actors and outside experts involved in adaptation planning need to recognise these knowledge systems and attempt to build a multi-cultural synergy out of diverse and plural knowledge systems. There has to be dialogue and deliberations on whose knowledge counts and why? We have to encourage the co-production of knowledge through partnerships between scientists and communities at risk (Berkes, 1999; Jackson, 2005). Such processes of knowledge inclusion would not only enhance equity and empowerment of marginalised communities but also develop sustainable adaptation strategies (Dixon, 2005; Srang-iam, 2013).

It is interesting to observe the interfaces emerging in the insurance sector as well. Crop insurance schemes to support farmers are often floated as an ideal adaptation option to deal with the impacts of climate change. However, when implemented in practice, it is observed that these schemes can make small and marginal farmers more vulnerable and poorer. The insurance companies treat each farmer claiming insurance as unique claimants. It is found that many claims may not get addressed; and many others have to wait without knowing when their claim would be addressed (Gupta, 2019). Studies show that usually less than 5 per cent of the claims are paid (Irshad, 2017). There is also a gap in which each individual farmer counts their crop loss and how the state and insurance companies decide upon the loss. Countries like Mongolia have launched an index-based livestock-insurance scheme for herders. It aims at the communal pooling of risk through promoting elements of self-insurance, market linkage and social obligations. There are different categorisations of risk, and insurances are calculated accordingly. Small losses are managed by the herders themselves. Large-scale losses are to be met by the insurance provider and any catastrophic loss will be managed by the government. Individual herders are to be paid when the mortality rates exceed local regional thresholds (Gandhi, 2018; World Bank, 2008). Though the scheme is innovative, the success of such a programme very much depends on the actor interfaces, and predominantly on who decides when the loss has crossed the local regional threshold. 'Whose knowledge of risk counts?' is a crucial determining question in such a context. If experts and bureaucrats are going to decide on such analysis, history shows that it will never benefit the people at the grassroots. Moreover, studies in India show that the only beneficiaries are the insurance companies, particularly private companies and not the vulnerable and marginalised farmers (Irshad, 2017, 2018). Similar interfaces are also found in the case of seed distribution by government departments to small and marginal farmers. The processes of identifying the beneficiaries are very cumbersome and bureaucratic. Verification of land records, tendering and bulk purchase of seeds often can be prolonged and ill-timed (Gandhi, 2018). The providers also lack reliable and updated information on seed varieties, price and possible pest risks that is simple and easy to understand (ibid). On many occasions, farmers end up receiving the seeds almost at the end of the cropping season. Often farmers are forced to depend on private seed producers and distributors of hybrid and transgenic seeds. Nevertheless, these firms are largely driven by the goals of profit maximization and often dispossess the farmers from the sovereign rights over their seeds and the produce. Farmers thus tend to be forced into a kind of dependency relationship with external stakeholders.

Several international humanitarian organisations have started innovative Community Cereal Banks (or lean-period cereal banks) across Africa to combat food insecurity. The cereal banks were established to address severe crises of food shortage at the community level. The banks are managed and run exclusively by women and provide poor farmers with access to cereal grains when there are seasonal or unexpected food shortages. The women-led management committee is responsible to undertake the whole process of identifying the vulnerable households, make arrangements for the storage of cereals, supply of essential food supplies to households, as well as ensure the replenishing of cereals on time. The stocks are replenished after the harvest on the basis of repayment in kind, with an interest rate negotiated with the beneficiaries that could go up to a maximum of 25 per cent. In parts of Niger, it was found that such innovations also helped in speedy agricultural recovery, reduced the distress sale of livestock, as well as prevented outmigration. The success stories of these banks have also led to its vast replication and scaling up (Hamadziripi, 2008). Nevertheless, recent studies show that these institutions are not viable in every context (Davis, 2015). The poor performance is largely attributed to inadequate deposits of cereals, stock decapitalisation caused by delays and non-payment of loans, lack of business acumen

on the part of the leaders and attitudes of members towards donor assistance (ibid). Studies also show that with scaling up and transitions in institutional arrangements, governance now is more complex and while
members still participate in decision-making, they are more dependent on
external actors for support in the management and governance (Davis,
2015; Hamadziripi, 2008).

The following section describes the emergence and flow of data in the adaptive
innovation process, and how these data could be analysed and interpreted.

Analysing data for reflective practice

Data analysis could be understood as the process associated with drawing
crucial meaning and understanding from the data collected during different
phases of the action research project (Rowley, 2014). Data analysis in adaptive
innovation has an important purpose. Similar to action research, it is used to
generate plans for further actions. More importantly, it has to be alert to the
subjugated voices of vulnerable groups with whom we work with and should be
able to locate specific practices that would enhance their capacities. The
interface analysis that evolves during each social encounter of the adaptive
innovation process should therefore provide the means of identifying those
vulnerable groups who may be otherwise excluded from particular situations of
practice. Our analysis should also focus on the nature of emerging conflicts and
incompatibility issues that could arise among diverse actors. Such an analytical
strategy also adds more meaning towards understanding the complexities and
dynamics involved in collective action and decision-making.

The data analysis in an actor-oriented approach involves examining the ways
in which different social actors manage and interpret ongoing and emergent
happenings in a particular social encounter or across diverse social encounters
(Long and Long, 1992). It involves a process of reflecting on how multiple
forms of knowledge–power relations are constituted and negotiated in these
social encounters (ibid). It therefore becomes crucial to analyse how diverse
actors respond differently to the problem, ideates to resolve it or evolve
appropriate mechanisms to implement these solutions. The differing actor
strategies and rationales, the conditions under which they arise, their viability or
effectiveness for solving specific problems and their structural outcomes; all
gain significance in the context of our analysis (Long, 1992).

In an actor-oriented approach, we entail a deep commitment to explore the
personal and shared lifeworlds of actors, the analysis of their actions and
decisions against the background of their lifeworlds (Seur, 1992). We have to
closely examine what actors in each social encounter are doing in the creation

and modification of diverse structural conditions that shape their actions (Villarreal, 1992). We also need to be aware of the recursive nature of these actions by being alert to how each actor perceives the decisions and actions of other actors (Seur, 1992). This also provides an understanding on the interlocking nature of social actions, propelled by divergent social interests, representations and consciousness (Long, 1992). More importantly, the complexities of layered social processes in terms of emergence, unintended consequences, system dynamics, adaptation and tipping points need to be analysed throughout our practice situations (Patton, 2011).

Data analysis should also aim at reflecting critically on the situatedness of self of the social worker with other community actors (Meerwald, 2013). In this regard, we must demonstrate reflexive awareness of the many factors that may have influenced our interpretations, judgements and decisions (Somekh, 1995). We in our practice encounters with other community actors should also reflect on how our interventions are contesting particular world views and normative behaviour. In particular, we should focus on whether we are imposing a particular normative framework of planned change on other actors. Moreover, we need to be cautious enough to reflect if our expertise, authority of knowledge and subsequent actions are disempowering our clients or impact groups (Hartman, 1994). In each social encounter, we have to identify the conditions under which particular definitions of reality are upheld by various actors; how these social construction of reality changes over time; and how factors such as insider–outsider relationships, our own positionality, perceptions, resources and investments shape these changes (refer to Figure 7.2). We will not only involve in a process of self-reflection of our own tacit knowledge-action frames but will also shape the emergence of on-the-spot variations in these knowledge and action frames, respectively. We have to understand that

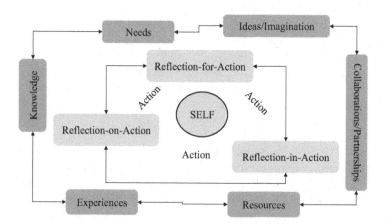

Figure 7.2 Reflective practice in adaptive innovation.

this involves a process of deconstructing our interventions as an ongoing, socially constructed and negotiated process among various actors (Long, 1992). In addition, we need to focus on our own dilemmas experienced in everyday negotiations and obligations with multiple actors.

Reflective note taking

Data captured through field notes, diaries and journals and memos form an important component of reflective inquiry. Field notes consist of our observations and notes on specific social encounters (Rodgers, 2008). They involve noting down the information gathered by the senses (Firmin, 2008). Reflective field notes can be written to analytically capture the process, findings, problems or patterns of the adaptive innovation process. It would also serve as a record of progress as well as a place to work out problems (Brodsky, 2008). Field notes also facilitate the discussions of priority issues emerging from the social encounters (Groenewald, 2008). Memos can be written along with field notes, and it largely involves the noting down of impressions, ideas, hunches and potential codes or themes (ibid). These can also be written down in diaries or journals. Diaries are personal reflective accounts situated around specific topics of interest or concern (Kemmis et al., 2014). A reflexive journal is more systematic than diaries but would help us to keep track of our thought processes during the adaptive innovation process (Kemmis et al., 2014; Rodgers, 2008). In practice, there may be considerable overlap among all of these forms of recording (ibid).

Field notes can be raw on-site notes and refined off-site notes. The raw on-site notes are also referred to as jottings that are made on a real-time basis (Emerson et al., 1995). The refined off-site notes or written-up field notes are a more cohesive write-up of one's jottings (Emerson et al., 1995; Nippert-Eng, 2015). The latter is written as soon as possible after a particular social encounter in the field to flesh out the rawer notes as accurately as possible (Nippert-Eng, 2015). We have to develop full field notes usually within a day of the events and social encounters while memory is still strong (McKechnie, 2008). While diaries and journals reflect our more personal statements, field notes are usually observations about social encounters involving other community actors. It is an enriching process to share our field notes to the community actors or other relevant members of the innovation platform, who are also involved in the co-creation of community-based climate change adaptation strategies. Field notes largely include written descriptions of social encounters. They may also contain maps, diagrams or other documents gathered from the sites of practice. The field notes are representations and interpretations of the social encounters in the field, which would help us to reconstruct our in-the-field experience (Nippert-Eng, 2015). They also help us to develop a better understanding of ourselves as a practitioner; and can be used to trace the action research journey (ibid). The noting down of descriptive details of social encounters as well as the simultaneous reflections on data, patterns and the process of research act as a form of quality control to the whole adaptive innovation process (Brodsky, 2008). Refined field notes are usually written as a first-person narrative of one's own musings and in an open, free-flowing, spontaneous manner (Brodsky, 2008; Kemmis et al., 2014). The field note could include the following aspects:

- The learning and action questions related to the specific social encounter.
- A briefing statement about the particular social encounter including both temporal and spatial dimensions.
- Description of the actors involved in the specific social encounter.
- A rough chronology of events that took place.
- A brief jotting down of actors' behaviour and verbal interactions including the verbatim, if possible, in the chronological sequence.
- Diagrammatic representations or mental models, if any.
- Reflections on the social encounter with respect to the learning and action questions.

The starting and ending point of recording each social encounter through field notes should be guided by the specific learning and action questions. Learning questions would vary for each phase of adaptive innovation and would help us to unearth the 'What', 'Why' and 'Where' of the innovation process. The action questions would specifically help us to reflect on 'then What', 'How' and 'When' of the planning and actions to be undertaken. If we lack such an agenda, there are chances that we could get a bit lost with the note taking as well as the fieldwork (Nippert-Eng, 2015). The focus of these recordings should always be on the emerging data and any important contextual constraints that both us and the community actors are getting into (ibid). Care has to be taken that the expanded field notes also document our biases, standpoints, dilemmas, possible mistakes, reactions and responses to the behaviour and interactions of community actors (Brodsky, 2008). They also could include our impressions, thoughts and feelings about each social encounter in the field (Owen, 2008).

Field notes can be written by hand using pen and paper or recorded using digital audio recorders (Angrosino, 2007). It is better to use the least intrusive form of recording such as noting down in a small notebook in the form of short phrases, abbreviations or mental models. These can be later transformed into full-scale descriptions with verbatim into a computer, but as quickly as possible following return from the field. Irrespective of the method of recording, what matters most is that we should be able to find and retrieve data as and when the need arises. It is a good idea to go through the whole set of refined field notes before proceeding with the more formal analysis and writing of adaptive innovation story. This would help in refreshing our memory and stimulate our reflection on our own knowledge and practice. Analysis of field notes could then be followed by thematically categorising its contents. Data analysis can be designed as a pathway for both on-site and off-site reflection. Both these forms of analysis could also feed into each other contributing to learning and meta learning (Rowley, 2014).

On-site analysis and reflection

On-site reflection happens simultaneously with the social encounters, as we examine the climate change issues and situations, nurture ideas and design actionable adaptation steps or while implementing the working models and pilots. During this phase of analysis, we would be largely involved in procedures of note taking and creating an assemblage of conversations that could be represented through certain relevant themes or codes. These strategies are aimed at discovering emerging patterns or meanings (experience/perception) within the data and organising them into a set of categories that signifies the experiences and perspectives of diverse actors (Stringer, 2007). The significant features and elements that make up the experience and perception of participant actors are identified during the on-site analysis.

We have to persistently uncover patterns of subordination or domination of specific groups of actors in the adaptive innovation process. The differences and contradictions across multiple voices and social positions of different actors constitute their situated knowledge. The differential patterns of these encounters can be understood only by examining the knowing of active subjects or actors including their interactions, negotiations and struggles in the social worlds (Knorr-Cetina, 1999; Long and Long, 1992). The responsibility therefore lies in closely reflecting upon the everyday life experiences and understandings of actors across intersectional dimensions of caste, class, gender, ethnicity, region, language or profession interacting in specific social encounters. The social construction of one's action and its relationship to continuity and change in one's situation of practice are also analytical elements in this context. The emphasis should be on to give voices to those subjugated narratives and knowledge systems through which actors imagine, relate and act in specific situations of practice.

Equally important is our ability to examine power relations in practice situations. Analysing power involves reflecting on how the values, interests, knowledge and actions of specific group of actors are being contested, ignored or recognised by other actors. A close examination of the social encounters would also unravel the nature of embedded power and its role in producing and sustaining inequalities among diverse actors in different situations of practice. It also helps us to reflect on the extent to which actors actually adhere to shared values and norms of their social world or examine the reasons why some actors deviate from the expected norms of behaviour. We need to simultaneously analyse how community actors judge the actions of others, the nature of their obligations and mutual commitments to act and reflect upon their actions. The situations, positions and discourses among actors must be analysed in detail (Clarke, 2005). Experiences of conflicts, cooperation and withdrawal/silence need to be reflected upon. Insights need to be gathered on how these processes among particular actors with specific interests were nurtured, sustained or altered. The analysis should reflect a kind of sensitivity and recognition to actor-specific knowledge systems and practices. The emphasis should be on the key experiences and events characterising certain sets of actions, their main features and relevant themes that could add meaning to our understanding of action.

Each phase of the adaptive innovation project would have generated its own sets of data. Field notes, journals, narratives and memos generated during each phase would remain as important sources of data for the on-site analysis. To begin with, we have to continuously review the nature and type of data that are being generated during each phase of the project. The focus here should be on the similarities and variations in the data and their relationship to the original concerns of inquiry that guided the intervention. An analysis of variance or differences could throw more light on emergence and discontinuities and could highlight the subjective contexts of the adaptive innovation processes. A closer exploration of the similarities could provide insights on the embedded nature of structures within the social ecological system including established patterns and order. Therefore, we have to look out for those sets of data that are directed to the initial lines of inquiry as well as cross-check for data that are deviating from these concerns. Perhaps these sets of data may highlight something more important and may be representing the voices of subjugated community actors. The next step is to sort diverse sets of data (ideas, concepts, events or experiences) into discrete units of meaning, which separate them from the rest of the information gathered (Stringer, 2007). Followed by which, assemble the sorted sets of data into specific themes and sub-themes emerging across diverse groups of actors. Once the thematic categories are identified, suitable mental models need to be evolved so as to design future course of action.

Off-site analysis and reflection

The off-site analysis happens at the end of each phase of the adaptive innovation project. These are largely reflexive processes, which explore the nature of interfaces, emergence, adaptation and coevolution during each phase of the adaptive innovation process. During this analytical phase, key experiences or transformational moments are selected out and analysed so as to illuminate the nature of those experiences (Stringer, 2007). In this phase, we seek to make sense and draw insights from the base of evidence and reflection that has emerged during the specific phases of adaptive innovation. These insights can also contribute to knowledge or theory (Rowley, 2014). The focus of off-site data analysis will be to explore the dynamics of actor interfaces during specific social encounters as well as to locate the reflective experiences of the self. The analysis could focus on how the spatial locations and social positions of community actors and outsiders shape the outcomes of the adaptive innovation project. The analysis could also look into the individual and collective agency of the vulnerable groups that we are working with. The practicalities underlying specific social encounters between different actors, demonstrating the interplay of diverse contextual factors that recursively shape human action also could be analysed. The analysis of the complexity and irregularity in actors' ways of imagination and action across diverse social encounters will also add value to reflective practice. It is also important to examine the processes by which ideologies are interpreted, manifested and enacted by diverse actors. We need to learn how ideas of vulnerable groups were created, nurtured and transformed during the adaptive innovation process. Significant insights can also be weaved into by analysing how ideas were adapted to local contexts and how they interact with locally predominant narratives (Appadurai, 1990; Sparke, 2004). Yet another important dimension is to examine how knowledge is co-created and sustained through such collaborative projects at the local level.

It is also important to examine how different social encounters are interactively connected through a complex network of interrelated actions of diverse actors. An understanding of how actions, interactions and practices are constituted across social-ecological systems is an essential representation of the adaptive innovation process. Such a strategy would also help us to understand the merits of those actions that were strategic in terms of how diverse actors agreed upon, negotiated or contested while arriving at decisions pertaining to the actions. The interface analysis gathers depth, when it is analysed across the temporal-spatial plane. Therefore, a cumulative interface analysis of the whole action research process needs to be carried out during the off-site data analysis phase. It needs to be examined in depth how actors with diverse values, interests, knowledge and power shaped the emergence and adaptation of the actionable processes. The off-site data analysis could also provide reflexive insights on how certain methods of inquiry were chosen, the spontaneity and relevance of those processes and how their selection proved to be crucial for the adaptive innovation process. The reflection can also be on what other pathways could have been chosen than the present one, and how different the outcomes would have been for the latter.

Our analysis will not be complete without self-reflection. This is also the essence of reflective practice. The transitions in our positionality, insider–outsider relationships and the roles, responsibilities and expectations vested with the professional self need to be analysed, interpreted and written about. Memos can facilitate reflective engagement during both on-site and off-site analysis. Memos are the best way to engage and reflect with one's own ideas, as they are sites of conversation with ourselves about our data (Clarke, 2005). They constitute an important analytical strategy as well. Memos are reflective assumptions (including personal, conceptual and theoretical ideas) on our actions and its consequences with respect to our understanding of each social encounter. They provide insights into the basic details of community actors' everyday practices revealing how the larger issues are inextricably interwoven with specific social encounters (Meerwald, 2013). Memos also aid in reflecting upon the nature of differential responses among community actors in each social encounter (Long, 1992). Early memoing happens during on-site analysis, when we are engaged in the field or when we begin to write field notes or transcribe interviews (van den Hoonaard and van den Hoonaard, 2008). These memos may involve our assumptions about what is going on with respect to each social encounter, or links to future learning and action questions. Memoing can also aid in the development of mental models and can be embedded within the field notes or written separately.

Today, there are several versions of software packages available for qualitative data analysis that lend themselves to the creation of diverse themes and memos, and network diagrams. However, it is recommended that the preliminary analysis should be done manually along with the participation of community actors. Flip charts, memo cards and diagrams could be used for the same. At the end of the off-site analysis, we should also be ready with the first draft of the adaptive innovation story. The draft story would present and detail the different phases of the project, the key actors and processes involved in each of these phases, the key outcomes and the relevant learning that emerged during the project.

References

Angrosino, M. (2007). *Doing ethnographic and observational research*, New Delhi: Sage.

Appadurai, A. (1990). Disjuncture and difference in the global cultural economy, in J.X. Inda and R. Rosaldo (Eds.). (2002). *The anthropology of globalization: a reader*, Oxford: Blackwell Publishing, pp. 46–64.

Arce, A., and Long, N. (1987). The dynamics of knowledge interfaces between Mexican agricultural bureaucrats and peasants: a case study from Jalisco, *Boletin de Estudios Latinoamericanos y del Caribe*, 43, pp. 5–30.

Berkes, F. (1999). *Sacred ecology: traditional ecological knowledge and resource management*, Philadelphia, PA: Taylor and Francis.

Brodsky, A.E. (2008). Fieldnotes, in L.M. Given (Ed.). *The Sage encyclopedia of qualitative research methods*, Vol. 1&2, New Delhi: Sage, pp. 341–343.

Clarke, A.E. (2005). *Situational analysis: grounded theory after the postmodern turn*, New Delhi: Sage.

Cleaver, F. (2001). Institutional bricolage, conflict and cooperation in Usangue, Tanzania, *IDS Bulletin*, 32(4), pp. 26–35.

Davis, M.G. (2015). The experience of community cereal banks in food – deficit areas of semi-arid Tanzania, *International Journal of Scientific and Research Publications*, 5(6), pp. 1–7.

de Wit, S. (2018). A clash of adaptations: how adaptation to climate change is translated in northern Tanzania, in S. Klepp and L. Chavez-Rodriguez (Eds.). *A critical approach to climate change adaptation: discourses, policies and practices*, London: Routledge-Earthscan, pp. 37–54.

Dixon, A. (2005). Wetland sustainability and the evolution of indigenous knowledge in Ethiopia, *The Geographical Journal*, 171, pp. 306–323.

Elsawah, S., Guillaume, J.H.A., Filatova, T., Rook, J., and Jakeman, A.J. (2015). A methodology for eliciting, representing, and analysing stakeholder knowledge for decision making on complex socio-ecological systems: from cognitive maps to agent-based models, *Journal of Environment Management*, 151, pp. 500–516.

Emerson, R.M., Fretz, R.I., and Shaw, L.L. (1995). *Writing ethnographic fieldnotes*, Chicago, IL and London: University of Chicago Press.

Firmin, M.W. (2008). Data collection, in L.M. Given (Ed.). *The Sage encyclopedia of qualitative research methods*, Vol. 1&2, New Delhi: Sage, pp. 190–192.

Gandhi, F.V. (2018). *A rural manifesto: realising India's future through her villages*, New Delhi: Rupa Publications.

Groenewald, T. (2008). Memos and memoing, in L.M. Given (Ed.). *The Sage encyclopedia of qualitative research methods*, Vol. 1&2, New Delhi: Sage, pp. 505–506.

Gupta, J. (2019). People struggle to survive India's drought, *India Climate Dialogue*, 17 June 2019. Retrieved from https://indiaclimatedialogue.net/2019/06/17/people-strug gle-to-survive-indias-drought/ [Last accessed on 21 July 2019].

Hamadziripi, A. (2008). Village savings and loans associations in Niger: Mata Masu Dubara model of remote outreach, case study, in N. Lee (Ed.). *Reaching the hard to reach: comparative study of member-owned financial institutions in remote rural areas*, Antigonish, Nova Scotia, Canada: Coady International Institute, pp. 1–30. Retrieved from https://coady.stfx.ca/wp-content/uploads/pdfs/ford/docs/Coady_Niger_Final.pdf [Last accessed on 9 January 2020].

Hartman, A. (1994). *Reflection and controversy: essays on social work*, Washington, DC: NASW Press.

Irshad, S.M. (2017). Agricultural risk and crop insurance in India: an analysis of public and private sector involvement in crop insurances, *European Insurance Law Review*, UDC, 368.5(540), pp. 46–53.

Irshad, S.M. (2018). The farmers' march should have its own agenda rather than Swami-nathan Commission, *Outlook*, 27 March 2018. Retrieved from www.outlookindia.com/ website/story/the-farmers-march-should-have-its-own-agenda-rather-than-swaminathan-commissions/310059 [Last accessed on 21 July 2019].

Jackson, P.R. (2005). Indigenous theorizing in a complex world, *Asian Journal of Social Psychology*, 8, pp. 51–64.

Kemmis, S., McTaggart, R., and Nixon, R. (2014). *The action research planner: doing critical participatory action research*, London: Springer.

Kirkwood, S., Jennings, B., Laurier, E., Cree, V., and Whyte, B. (2016). Towards an inter-actional approach to reflective practice in social work, *European Journal of Social Work*, 19(3–4), pp. 484–499.

Klepp, S., and Chavez-Rodriguez, L. (2018). Governing climate change: the power of adaptation discourses, policies, and practices, in S. Klepp and L. Chavez-Rodriguez (Eds.). *A critical approach to climate change adaptation: discourses, policies and practices*, London: Routledge-Earthscan, pp. 3–34.

Knorr-Cetina, K. (1999). *Epistemic cultures: how the sciences make knowledge*, Cambridge, MA: Harvard University Press.

Long, N. (1989). Conclusion: theoretical reflections on actor, structure and inter-face, in N. Long (Ed.). *Encounters at the interface: a perspective on social discontinuities in rural development*, Wageningen: Agricultural University Wageningen, pp. 221–243.

Long, N. (1992). From paradigm lost to paradigm regained: the case for an actor-oriented sociology of development, in N. Long and A. Long (Eds.). *Battlefields of knowledge: the interlocking of theory and practice in social research and development*, London: Routledge, pp. 16–43.

Long, N. (2001). *Development sociology: actor perspectives*, London: Routledge.

Long, N., and Long, A. (Eds.). (1992). *Battlefields of knowledge: the interlocking of theory and practice in social research and development*, London: Routledge.

McKechnie, L.E.F. (2008). Observational research, in L.M. Given (Ed.). *The Sage encyclopedia of qualitative research methods*, Vol. 1&2, New Delhi: Sage, pp. 573–576.

Meerwald, A.M.L. (2013). Researcher/Researched: repositioning research paradigms, *Higher Education Research and Development*, 32(1), pp. 43–55.

Morchain, D. (2018). Rethinking the framing of climate change adaptation: knowledge, power, and politics, in S. Klepp and L. Chavez-Rodriguez (Eds.). *A critical approach to climate change adaptation: discourses, policies and practices*, London: Routledge-Earthscan, pp. 55–73.

Nippert-Eng, C. (2015). *Watching closely: a guide to ethnographic observation*, New York: Oxford University Press.

Owen, J.A.T. (2008). Naturalistic inquiry, in L.M. Given (Ed.). *The Sage encyclopedia of qualitative research methods*, Vol. 1&2, New Delhi: Sage, pp. 547–550.

Papell, C.P. (1996). Reflections on issues in social work education, in N. Gould and I. Taylor (Eds.). *Reflective learning for social work*, Brookfield, VT: Ashgate, pp. 11–22.

Patton, M.Q. (2011). *Developmental evaluation: applying complexity concepts to enhance innovation and use*, New York: The Guilford Press.

Rodgers, B.L. (2008). Audit trail, in L.M. Given (Ed.). *The Sage encyclopedia of qualitative research methods*, Vol. 1&2, New Delhi: Sage, p. 43.

Rowley, J. (2014). Data analysis, in D. Coghlan and M. Brydon-Miller (Eds.). *The Sage encyclopedia of action research*, New Delhi: Sage, pp. 239–242.

Santha, S.D. (2015). Early warning systems among the coastal fishing communities in Kerala: a governmentality perspective, *Indian Journal of Social Work*, 76(2), pp. 199–222.

Schön, D.A. (1985). *The design studio: an exploration of its traditions and potentials*, London: RIBA Publications.

Scott, J.C. (1985). *Weapons of the weak: everyday forms of peasant resistance*, New Haven, CT: Yale University Press.

Seur, H. (1992). The engagement of researchers and local actors in the construction of case studies and research themes, in N.E. Long and A. Long (Eds.). *Battlefields of knowledge: the interlocking of theory and practice in social research and development*, London: Routledge, pp. 115–143.

Somekh, B. (1995). The contribution of action research to development in social endeavours: a position paper on action research methodology, *British Educational Research Journal*, 21(3), pp. 339–355.

Sparke, M. (2004). Political geographies of globalization: (1) dominance, *Progress in Human Geography*, 28, pp. 777–794.

Srang-iam, W. (2013). De-contextualized knowledge situated politics: the new scientific–local politics of rice genetic resources in Thailand, *Development and Change*, 44(1), pp. 1–27.

Stringer, E.T. (2007). *Action research*, New Delhi: Sage.

Ulloa, A. (2018). Reconfiguring climate change adaptation policy: indigenous people's strategies and policies for managing environmental transformations in Colombia, in S. Klepp and L. Chavez-Rodriguez (Eds.). *A critical approach to climate change adaptation: discourses, policies and practices*, London: Routledge-Earthscan, pp. 222–237.

van den Hoonaard, D.K., and van den Hoonaard, W.C. (2008). Data analysis, in L.M. Given (Ed.). *The Sage encyclopedia of qualitative research methods*, Vol. 1&2, New Delhi: Sage, pp. 186–188.

Villarreal, M. (1992). The poverty of practice: power, gender and intervention from an actor-oriented perspective, in N.E. Long and A. Long (Eds.). *Battlefields of knowledge: the interlocking of theory and practice in social research and development*, London: Routledge, pp. 247–267.

Viterna, J. (2013). *Women in war: the micro-processes of mobilization in El Salvador*, New Delhi: Oxford University Press.

World Bank. (2008). *World development report: agriculture for development*, Washington, DC: The World Bank.

8   Concluding reflections

The aim of this book was to introduce and elaborate a community-based innovation model to design and practice climate change adaptation strategies. This work was undertaken with the realisation that social workers as reflective practitioners could play an important role in enabling vulnerable populations to cope with climate change and develop opportunities for sustainable development. Rooted in action research, this model could enable us social workers and other development practitioners to facilitate the evolution of people-centred ethical adaptation mechanisms to climate change. Adaptive innovation aims at vulnerability reduction and strengthening the adaptive capacities of social–ecological systems. It nurtures values of justice, care and solidarity to stimulate shared conversations and co-creation of adaptation solutions. It is an enduring journey where diverse actors navigate collectively to make their lived environment safe and secure, mutually learn and make choices and create meaning out of these social encounters.

A resilient natural environment in itself is a meta-capability that enables all other capabilities in a person to live a dignified life (Holland, 2012). A stable climate system should then be understood as one component of a broader environmental meta-capability that supports and enables all other central human capabilities (ibid). As social workers, our persistent efforts towards raising awareness and promoting advocacy campaigns to mitigate environmental problems have achieved considerable recognition at the global level. During the 1990s, our contributions to environmental justice were recognised by international organisations such as the United Nations and the Right Livelihood Foundation. However, today in the context of global warming and climate change, the practice of environmental social work has become more complex. Social workers involved in climate change adaptation, sustainable livelihoods or disaster risk reduction are repeatedly facing several challenges in their practice fields due to the vagaries and unpredictability of the natural environment. Their struggles vary from restoring fragile ecosystems and promoting alternative livelihoods to saving lives or relocating people whose properties and infrastructure have been damaged. Enhancing the adaptive capacities of vulnerable populations to climate change therefore become crucial in our everyday practice.

We have to strive towards new forms of knowledge production and action-oriented learning, which also involves considerable reflection on how different community actors think and act in complex situations and what its implications are (Dominelli et al., 2018; Taylor, 1996). Therefore, our methods of inquiry and practice will require continuous reflection, innovation and improvisation. Recognition of localised, people-centred knowledge systems and decision-making are crucial in this context. For instance, the recent Lancet Commission Report on 'the global syndemic of obesity, undernutrition and climate change' insists that reviving traditional knowledge systems could be a significant step towards conserving biodiversity and ensuring food security (Swinburn et al., 2019). The report asserts that 'bringing indigenous and traditional knowledge to this effort will also be important because this knowledge is often based on principles of environmental stewardship, collective responsibilities, and the interconnectedness of people with their environments' (Swinburn et al., 2019, p. 794).

Adaptation decisions has to be based on what the people need, think and know, rather than following dogmatic, top-down representations. Adaptation can mean different things to diverse actors and specifically for those in power, it can be strategized as a means to further their own interests (de Wit, 2018). Certain community-based adaptation strategies that are prevalent today seek to impart a kind of mechanical participation among local community actors. And at the same they tend to favour a few elite actors in the society (Gotham, 2012; Tierney, 2015). For example, a sea wall can be a reactionary and individualistic approach to the problem of sea-level rise; and such technocentric strategies may destroy the confidence and motivation of local community actors to participate in adaptation projects (Hoggett et al., 2009; Uncapher and Yvellez, 2019). Ethical adaptation demands that we at least change such predominant ideas and resist the strategisations and pursuance of certain vested interests (Bendik-Keymer, 2012). We have to always remain alert to such interests and power dynamics. The ideological underpinning behind every adaptation decision has to be therefore analysed critically and reflectively. In addition to the exploration of who participates in adaptation decision-making and action framing, it is also equally important to examine the justice implications of adaptation projects in terms of their outcomes (Morchain, 2018).

Any adaptation decision has to take into account the situatedness and contextual positioning of vulnerable community actors. Intersectional factors such as gender, race, caste, class, educational attainment, sexual orientation, disability status, age and other aspects of their identity could influence decision-making in the adaptation process (Brydon-Miller, 2008). All vulnerable groups may not have the agency to voice their state of vulnerability and powerlessness (Duffy, 2008). In this context, privileged actors in positions of power may not always give informed consent to certain decisions, as their own particular interests and power would be at stake (Duffy, 2008; Foster and Glass, 2017). There are also possibilities of elite capture of opportunities arising out of the very conceptualisation of adaptation and excluding the poor and marginalised from such spaces (de Wit, 2018). The success of any adaptation project will depend on how sensitive we and other actors to these issues of inclusion and exclusion are and how are we responding to the fears and resentments of the marginalised groups (Parrott, 2010).

For adaptation to be innovative and inclusive, we have to understand and construe the contexts of adaptation in all its complexity and knowledge plurality (Morchain, 2018). A failure to do so could risk disempowering the very groups that an adaptation initiative sought to support (ibid). We need to be aware and sensitive to the process of othering that could happen even in adaptation contexts, where the poor and marginalised can be discriminated and denied access to basic livelihood resources and decision-making structures (Parrott, 2010). We have to thus reflexively engage with diverse ethical complexities and maintain a sense of openness to our practice situation as well (Foster and Glass, 2017). This is in itself a challenge to the whole climate change adaptation project. In addition, the emergent and uncertain nature of adaptation outcomes could make our practice situations quite uneven and difficult (Kronlid, 2014; Schneider and Lane, 2006).

The adaptive innovation model offers scope to develop more open and deliberative alternatives, when compared to previous top-down strategies of adaptation. It also voices for an explicit recognition of the situatedness of actors and the plurality of knowledge systems and interests. And to be specific, the emphasis is towards recognizing the knowledge possessed by marginalised communities and vulnerable groups and ensuring their representation in the adaptation processes. The ultimate outcome of such an approach will be that community actors with whom we work are able to take control over the forces that shape their lives (Edwards, 1989; Hetherington and Boddy, 2013). Our aim has to be towards enhancing the capacities of local community actors to learn, experiment and reorganise to the emerging threats and changing surroundings (Aase et al., 2013; Krasny et al., 2011; Reed et al., 2015). In this regard, we have a crucial role in revitalising a culture of reflective learning and nurturing an iterative process of planning, action and contemplation.

What constitutes an actual innovation in adaptation can only be determined on the basis of its practical application and use (Ellström, 2010). The ultimate outcome being that all involved actors are able to develop their individual and collective capacities to solve complex adaptation challenges. The innovation environment should stimulate shared conversations and build solidarity networks that are meaningful, proactive, balanced, responsible and have faith in one another. Solidarity implies that we 'stretch ourselves to connect with the other' (Jobin-Leeds and AgitArte, 2016, p. 427). During crisis situations, people not only re-enact everyday patterns of solidarity, but also renegotiate and establish new forms of expectations, obligations and relationships between them (Reyes, 2016). The innovation environment should look at their diverse perspectives with respect and sustain a commitment to recognise each actor's experience, interests, emotions and values. The strength and direction of actors' emotions and the values they uphold reveal the meaning, importance and ethical concerns that they attribute to the practice situation in which the innovation is being tested or implemented (Wagenaar and Cook, 2003).

Informed consent is foundational to the whole adaptive innovation process. We are ethically bound to protect participant actors from maltreatment by requiring us and other external experts to respect their autonomy and secure their active agreement to participate in the project (Foster and Glass, 2017). We should view our ethical responsibilities as only partially over when we receive informed consent from participant actors at the beginning of the project (Owen, 2008). The consent-seeking process has to be understood as dynamic and ongoing throughout the life of the adaptive innovation project. Revisiting the issue of consent throughout the project is essential for us to maintain the trust and commitment of participant actors and relationships that develop in the field (ibid). We also need to ensure that the consent seeking happens in a form and context, which is accessible to all participants (Adams et al., 2015). The progress of work should remain fully visible to all participants (Winter and Munn-Giddings, 2001). All the discussions need to be fully documented and has to be made available to those who were not present. We have the responsibility to inform participant actors about the emerging developments including certain key findings, interpretations and decisions taken at each phase of the adaptive innovation project. The recording and documentation procedures have to be agreed in advance after discussion and taking consent with participant actors (Kemmis and McTaggart, 2000; Winter and Munn-Giddings, 2001).

As some of the adaptive innovation methods of inquiry rely on visual narratives, there are certain pervasive ethical issues that could emerge during these processes as well. Images and photographs may not hold a steady fixed meaning (Bach, 2008). We and other participant actors should be aware of the temporality of visual narratives and the many meanings it could convey over a period of time involving stories of both the present and the reconstructed past (ibid). Similar to photographs, maps are also not neutral objects that could be separated from the social context. Therefore, ethical reflection surrounding the power of visual narratives and methodologies such as photographs and maps becomes very important (Barkved et al., 2014). There needs to be a sense of relational ethics in the writing and documentation of adaptation projects as well. We should avoid blaming and shaming when narrating experiences associated with certain social encounters (Adams et al., 2015). It is also our responsibility to tell stories from the field that are both faithful to experience and respectful of relationships. A participant should be given the complete right to amend her or his contribution before it is circulated to others (Kemmis and McTaggart, 2000; Whitehead and McNiff, 2006; Winter and Munn-Giddings, 2001). Also, there has to be a clear distinction between documents that are confidential to project participants and reports intended for wider publication (ibid).

Exchange and learning thus remain central to the adaptive innovation process. Participation in such contexts requires a curious mind and a willingness to work in contexts of uncertainty and openness (Southern, 2015). Participant actors should be able to steer the project from being an external-led initiative to a self-organised platform (Ngwenya and Hagmann, 2011). It is important that community actors who are active members of the innovation platform actively participate and feel committed and maintain a sense of ownership of the whole process. Past successes and failures of community actors in adaptation have to be recognised and respected. We should equally be aware that actors know not only about their choices but also the costs of these decisions. In this regard, the decisions of local community actors to not to adopt innovations, to postpone embracing an innovation or to discontinue the use of an innovation has to be valued (Beckford, 2002).

As social workers, we need to be aware about the educational and reflective nature of our practices (Horton and Freire, 1990). The reflective learning happens before, during and after every social encounter and is linked to our very purpose of existence and curiosity (ibid). As social workers, we are an outsider-within throughout the adaptive innovation process. Unlike a detached outsider, the outsider-within denotes engagement and interest rather than disinterest. We are both a participant and an observer; being part of a social group and yet we are separated from it because of our professional values and practice skills. Often, positioning ourselves as the situated-other, we could develop alternative methodologies that would involve the active participation of community actors in the investigation and transformation of their present social reality (Edwards, 1989; Lykes, 1997). At the same time, we have to be aware of our own subjectivity and how it is shaped by the participant actors' thinking and doing. The participant actors themselves could engage in a process of self-reflection and attempt to let go of certain constraining feelings that act as barriers to innovation (Gilpil-Jackson, 2015). In this regard, listening to people's voice is an important skill.

Environmental crisis such as climate change can be considered as opportunities to innovate appropriate solutions to sustain the ecological and livelihood security of a community. Our planning processes need to be flexible enough to deal with the emergent nature of adaptive innovation. Initial successes could generate temporary commitments among participant actors. As community actors continuously strive towards dealing with uncertainty, surprises or crisis, we can trust that they themselves have reinvigorated a journey to strengthen the resilience of their social–ecological systems. Nevertheless, engaging with actors, developing relationships, mobilising resources and bringing about long-lasting change takes time. Nurturing, strengthening and sustaining the voices, aspirations, ideas, imaginations and hopes of community actors therefore becomes crucial throughout this journey. As a parting reflection, I would like to quote the words of Jobin-Leeds and AgitArte (2016, p. 439),

> We will fail. We will stumble. We will doubt and we will hesitate. Don't give up. Know that we are making a difference. Even if we make mistakes, we are testing out what is possible in our moment in history … Together, we can build a world full of love, awareness, critical reflection, creativity, humanity, understanding, and meaning, bringing out each other's best.

References

Aase, T.H., Chapagain, P.S., and Tiwari, P.C. (2013). Innovation as an expression of adaptive capacity to change in Himalayan farming, *Mountain Research and Development*, 33, pp. 4–10.

Adams, T.E., Jones, S.H., and Ellis, C. (2015). *Autoethnography*, New York: Oxford University Press.

Bach, H. (2008). Visual narrative inquiry, in L.M. Given (Ed.). *The Sage encyclopedia of qualitative research methods*, Vol. 1&2, New Delhi: Sage, pp. 938–940.

Barkved, L., de Bruin, K., and Romstad, B. (2014). *Mapping of drought vulnerability and risk*, Final report on WP 2.3: Extreme Risks, Vulnerabilities and Community based-Adaptation in India (EVA): A Pilot Study, New Delhi: CIENS-TERI, TERI Press.

Beckford, C.L. (2002). Decision–making and innovation among small-scale yam farmers in central Jamaica: a dynamic, pragmatic and adaptive process, *Royal Geographical Society*, 168(3), pp. 248–259.

Bendik-Keymer, J. (2012). Ethical adaptation to climate change: beyond 'a safe operating space for humanity', *The European Financial Review*, 3 October 2012. Retrieved from www.europeanfinancialreview.com/ethical-adaptation-to-climate-change/ [Last accessed on 17 July 2019].

Brydon-Miller, M. (2008). Ethics and action research: deepening our commitment to principles of social justice and redefining systems of democratic practice, in P. Reason and H. Bradbury (Eds.). *The Sage handbook of action research: participative inquiry and practice*, 2nd edition, New Delhi: Sage, pp. 199–210.

de Wit, S. (2018). A clash of adaptations: how adaptation to climate change is translated in northern Tanzania, in S. Klepp and L. Chavez-Rodriguez (Eds.). *A critical approach to climate change adaptation: discourses, policies and practices*, London: Routledge-Earthscan, pp. 37–54.

Dominelli, L., Nikku, B.R., and Ku, H.B. (2018). Introduction: why green social work? in L. Dominelli (Ed.). *The Routledge handbook of green social work*, London: Routledge, pp. 1–6.

Duffy, M. (2008). Vulnerability, in L.M. Given (Ed.). *The Sage encyclopedia of qualitative research methods*, Vol. 1&2, New Delhi: Sage, pp. 944–948.

Edwards, M. (1989). The irrelevance of development studies, *Third World Quarterly*, 11(1), pp. 116–135.

Ellström, P. (2010). Practice-based innovation: a learning perspective, *Journal of Workplace Learning*, 22(1/2), pp. 27–40.

Foster, S.S., and Glass, R.D. (2017). Ethical, epistemic, and political issues in equity-oriented collaborative community-based research, in L.L. Rowell, C.D. Bruce, J.M. Shosh, and M.M. Riel (Eds.). *The Palgrave international handbook of action research*, New York: Palgrave Macmillan, pp. 511–525.

Gilpil-Jackson, Y. (2015). Transformative learning during dialogic organisation development, in G.R. Bushe and R.J. Marshak (Eds.). *Dialogic organisation development: the theory and practice of transformational change*, Oakland, CA: Berrett-Koehler Publishers, pp. 245–267.

Gotham, K.F. (2012). Disaster inc: privatization and post-Katrina rebuilding in New Orleans, *Perspectives on Politics*, 10(3), pp. 633–646.

Hetherington, T., and Boddy, J. (2013). Ecosocial work with marginalised populations: time for action on climate change, in M. Gray, J. Coates, and T. Hetherington (Eds.). *Environmental social work*, London: Routledge, pp. 46–61.

Hoggett, P., Mayo, M., and Miller, C. (2009). *The dilemmas of development work: ethical challenges in regeneration*, Bristol: The Policy Press.

Holland, B. (2012). Environment as meta-capability: why dignified human life requires a stable climate system, in A. Thompson and J. Bendik-Keymer (Eds.). *Ethical adaptation to climate change: human virtues of the future*, Cambridge, MA: The MIT Press, pp. 145–164.

Horton, M., and Freire, P. (1990). *We make the road by walking: conversations on education and social change*, Philadelphia, PA: Temple University Press.

Jobin-Leeds, G., and AgitArte. (2016). Epilogue: solidarity – a gathering, in G. Jobin-Leeds and AgitArte (Ed.). *When we fight, we win: twenty-first-century social movements and the activists that are transforming our world*, e-book, New York: The New Press, pp. 421–440. ISBN 978-1-62097-140-6.

Kemmis, S., and McTaggart, R. (2000). Participatory action research, in N.K. Denzin and Y. S. Lincoln (Eds.). *The SAGE handbook of qualitative research*, London: Sage, pp. 567–606.

Krasny, M.E., Lundholm, C., and Plummer, R. (Eds.). (2011). *Resilience in social-ecological systems: the role of learning and education*, Abingdon: Routledge.

Kronlid, D.O. (2014). *Climate change adaptation and human capabilities: justice and ethics in research and policies*, New York: Palgrave Macmillan.

Lykes, M.B. (1997). Activist participatory research among the Maya of Guatemala: constructing meanings from situated knowledge, *Journal of Social Issues*, 53(4), pp. 725–746.

Morchain, D. (2018). Rethinking the framing of climate change adaptation: knowledge, power, and politics, in S. Klepp and L. Chavez-Rodriguez (Eds.). *A critical approach to climate change adaptation: discourses, policies and practices*, London: Routledge-Earthscan, pp. 55–73.

Ngwenya, H., and Hagmann, J. (2011). Making innovation systems work in practice: experiences in integrating innovation, social learning and knowledge in innovation platforms, *Knowledge Management for Development Journal*, 7(1), pp. 109–124.

Owen, J.A.T. (2008). Naturalistic inquiry, in L.M. Given (Ed.). *The Sage encyclopedia of qualitative research methods*, Vol. 1&2, New Delhi: Sage, pp. 547–550.

Parrott, L. (2010). *Values and ethics in social work practice*, Exeter: Learning Matters Ltd.

Reed, S.O., Friend, R., Jarvie, J., Henceroth, J., Thinphanga, P., Singh, D., Tran, P., and Sutarto, R. (2015). Resilience projects as experiments: implementing climate change resilience in Asian cities, *Climate and Development*, 7(5), pp. 469–480.

Reyes, J.A.C. (2016). *Disaster citizenship: survivors, solidarity, and power in the progressive era*, Chicago, IL: University of Illinois Press.

Schneider, S.H., and Lane, J. (2006). Dangers and thresholds in climate change and the implications for justice, in A.W. Neil, J. Paavola, S. Huq, and M.J. Mace (Eds.). *Fairness in adaptation to climate change*, Cambridge, MA: The MIT Press, pp. 23–52.

Southern, N. (2015). Framing inquiry: the art of engaging great questions, in G.R. Bushe and R.J. Marshak (Eds.). *Dialogic organisation development: the theory and practice of transformational change*, Oakland, CA: Berrett-Koehler Publishers, pp. 269–289.

Swinburn, B.A., Kraak, V.I., Allender, S., Atkins, V.J., Baker, P.I., Bogard, J.R., Brinsden, H., Calvillo, A., De Schutter, O., Devarajan, R., Ezzati, M., Friel, S., Goenka, S., Hammond, R.A., Hastings, G., Hawkes, C., Herrero, M., Hovmand, P.S., Howden, M., Jaacks, L.M., Kapetanaki, A.B., Kasman, M., Kuhnlein, H.V., Kumanyika, S.K., Larijani, B., Lobstein, T., Long, M.W., Matsudo, V.K.R., Mills, S.D.H., Morgan, G., Morshed, A., Nece, P.M., Pan, A., Patterson, D.W., Sacks, G., Shekar, M., Simmons, G.L., Smit, W., Tootee, A., Vandevijvere, S., Waterlander, W.E., Wolfenden, L., and Dietz, W.H. (2019).

The global syndemic of obesity, undernutrition, and climate change: the Lancet Commission Report, *The Lancet*, 393(10173), pp. 791–846.

Taylor, I. (1996). Reflective learning, social work education and practice in the 21st century, in N.T. Gould and I. Hants (Eds.). *Reflective learning for social work*, Burlington: Ashgate Publishers Ltd, pp. 153–161.

Tierney, K. (2015). Resilience and the neoliberal project: discourses, critiques, practices – and Katrina, *American Behavioral Scientist*, 59(10), pp. 1327–1342.

Uncapher, D., and Yvellez, C. (2019). Climate adaptation needs to put human rights above property values, *Truthout*, 14 July 2019. Retrieved from https://truthout.org/articles/climate-adaptation-needs-to-put-human-rights-above-property-values/ [Last accessed on 16 July 2019].

Wagenaar, H., and Cook, N. (2003). Understanding policy practises, in M. Hajer and H. Wagenaar (Eds.). *Deliberative policy analyses: understanding governance in the network society*, Cambridge: Cambridge University Press, pp. 139–171.

Whitehead, J., and McNiff, J. (2006). *Action research living theory*, New Delhi: Sage.

Winter, R., and Munn-Giddings, C. (2001). *A Handbook for action research in health and social care*, London: Routledge.

Glossary

Action Framing A collective and participatory process that involves translating the emergent ideas into meaningful sensory experiences, images or visuals of action.

Action Research A transformative change-making approach guided by the values of justice, care and solidarity that are constituted through participatory inquiry and reflective practice.

Actor Interfaces Critical points of interaction located in specific social encounters that involve multiple actors with diverse values, interests, knowledge and power; and the consequences could be either social innovations or social discontinuities.

Actors Individuals, groups and organisations within a specific social–ecological system having a set of unique values, commitments, ideologies and interests; and are recognised as active stakeholders with specific knowledge and capacity for action.

Adaptation The processes, actions and outcomes that could enable social–ecological systems or its elements to better cope with, manage or adjust to actual or expected climate and its effects.

Adaptive Innovation Refers to people-centred innovation processes by which local community actors collectively analyse their own situations in the context of social and ecological transitions; forge constructive partnership with other relevant actors to dialogue, ideate and develop working models; and implement and critically observe, reflect and validate their adaptive strategies to the emergent contexts. These processes are situated, reflective, context specific, developmental and committed to the values of care, justice and solidarity.

Ashram A spiritual abode where one strives to lead a self-seeking and disciplined life.

Burra Katha A traditional oral storytelling practice.

Chauka A water-harvesting structure in the shape of a square open on one side and surrounded by earthen bunds of varying heights.

Desaria Dhan A deep-water rice variety.

Dialogic Ideation A process of collective imagination where community actors in partnership with other stakeholders attempt to ideate and co-create multiple adaptation pathways through in-depth deliberations, dialogue and other forms of shared conversations and decision-making.

Emergence The emerging novel and self-organised patterns of systemic arrangements, which represent both the intended and unintended consequences of the adaptation project.

Gram Panchayat A form of local self-governance institution at the village level in India.

Hari Katha A traditional oral storytelling practice.

Innovation Platforms A collective and collaborative institutional space owned and governed by community actors in partnership with other relevant stakeholders to mutually share knowledge, imagine and innovate situated practices that could reduce vulnerabilities and build resilience of social–ecological systems.

Micro-mobilisation A strategic process aimed at organising community actors and other relevant stakeholders to participate collectively through innovation platforms in developing suitable adaptation strategies.

Piloting An iterative and reflective process to implement and test the suitability, feasibility and effectiveness of working models.

Reflective Practice Iterative cycles of action-reflection, where reflection on action taken at each phase of adaptive innovation would set the stage for the next phase.

Sarvodaya The Gandhian principle that advocates the universal well-being of all human and other sentient beings.

Satyagraha The Gandhian principle that insists on non-violent action to bring about transformative change by holding on to the truth.

Shramdaan A voluntary contribution of labour, often carried out for a social, political or ecological cause.

Situated Knowledge Subjective knowledge that is local, partial, plural and has its value on the particular situation at hand, which is embedded with the social positions and lived experiences of actors.

Situational Analysis The process of understanding the vulnerability contexts, livelihood practices, adaptation trends and other key issues affecting diverse community actors in a given social–ecological system.

Social–ecological systems An integrated perspective of humans-in-nature, which recognises that humans and ecosystems are intricately interconnected, each affecting the other and often in complex ways.

Vulnerability The intrinsic characteristics and manifested conditions of humans and non-humans in a social-ecological system that could constrain their capacity to cope with, manage or adjust to actual or expected climate and its effects.

Index